Crystal Research and Technology

8

Journal of Experimental and Industrial Crystallography

Zeitschrift für experimentelle und technische Kristallographie

Volume 22
1987
Number 8 August

Akademie-Verlag · Berlin

ISSN 0232-1300 Cryst. Res. Technol., Berlin 22 (1987) 8, 987–1114, K125–K140

Directions to the author

In the journal "Crystal Research and Technology – Kristall und Technik" as a rule the following categories of papers will be printed: original papers, review articles, short notes, and book reviews, preferentially in **English**. Complete information necessary for type-setting and a "Short Title" should be added.

The manuscript should not exceed 10 pages and be type-written on one side only in the format A4 (21 × 29.7 cm) on white ink-proof paper, with a two-lined distance and a left-hand free edge of 3 cm, and submitted in 2 copies. All sheets of the paper belonging to the contribution are continuously to be numbered.

The head of the manuscript should be arranged in this manner: initials and name of the author; institution, and title of the paper. Then two abstracts should follow, one in the original language and the second in German, English, French, or Russian, respectively.

Formulae, signs and symbols, letters etc. being wanting on the type-writer, should be distinctly written with penholder. O as zero is to be underlined, the capital O is not. Greek letters should be underlined with red ink.

Paragraphs should be marked by indenting of at least 3 touches.

Concerning the marking of the manuscript should be noted:
headings should be underlined, sub-headings not, and indented by 9 touches. The titles of tables are not to be underlined. Other markings in the text will be done by the editorial staff. They mean:

light-red ~~~~~~~~ = small capitals
blue ~~~~~~~~ = in italics

light-red | perpendicular straight red line on the left edge = small printing

Tables are to be added on separate sheets of paper at the end of the contribution.

Foot notes should be continuously written on a separate sheet of paper.

Figures (Fig.) – drawings and photographs – should be not more than 8 and submitted (2 copies) in a special envelope: photographs as white high-gloss prints (inscription on a transparent covering sheet) and drawings as distinct sketches. On the back of the submitted figures the reduction scale in tenth (9/10, 8/10, 5/10 e.g.), author's name, short title of the paper, and the number of the figure should be given.

Captions for the figures should be submitted on a separate sheet of paper.

References should be written without numbers in alphabetical order of the names at the end of the contribution. Cited names in the text stand for the references. The references should contain the following items, for example:

for books:	RAST, K.: Physikalisch-chemische Rechnungen, Berlin 1958
for journals:	KRAUSE, H. N., MÜLLER, K.: Cryst. Res. Technol. **16**, 70 (1981)

Add, please, the short title, about 6 keywords, and the academic grades of the authors.
The authors will be sent two galley proofs of review articles and original papers. Proof reading of short notes and book reviews will be done by the editorial staff.

For each of these papers a total of 50 reprints is available free of charge. Requests for additional maximum 200 reprints (against payment) should be given together with the submitted manuscript.

8

Crystal Research and Technology

Founded 1966 as
Kristall und Technik

Journal of Experimental and Industrial Crystallography

ESTABLISHED BY
W. KLEBER and H. NEELS

EDITED BY
H. NEELS, Leipzig

EDITOR-IN-CHIEF
H. NEELS, Leipzig

VOLUME 22 · 1987 · NUMBER 8

THEORIA CUM PRAXI

Akademie-Verlag · Berlin

Papers should be directed to the editor's office
DDR-7030 Leipzig, Scharnhorststraße 20,
or to the editor Prof. Dr. Hermann Neels,
DDR-7254 Machern, Zeititzer Weg 25,
or to one of the members of the Advisory Board.

Orders can be sent

- in the GDR:
  to the Postzeitungsvertrieb or to the Akademie-Verlag Berlin, Leipziger Straße 3 – 4, DDR-1086 Berlin;
- in the other socialist countries: to a book-shop for foreign language literature or to the competent news-distributing agency;
- in the FRG and Berlin (West): to a book-shop or to the wholesale distributing agency Kunst und Wissen, Erich Bieber OHG, Wilhelmstr. 4 – 6, D-7000 Stuttgart 1;
- in the other Western European countries: to Kunst und Wissen, Erich Bieber GmbH, General Wille-Str. 4, CH-8002 Zürich;
- in USA and Canada to the VCH Publishers, Inc., 303 NW 12th Avenue, Deerfield Beach, FL 33442-1705, USA;
- in other countries; to the international book- and journal-selling trade, to Buchexport, Volkseigener Außenhandelsbetrieb der DDR, Postfach 160, DDR-7010 Leipzig, or to the Akademie-Verlag Berlin, Leipziger Straße 3 – 4, DDR-1086 Berlin.

Journal „Crystal Research and Technology"

Herausgeber: Im Auftrage des Akademie-Verlages Berlin herausgegeben von Prof. Dr. Hermann Neels.

Verlag: Akademie-Verlag Berlin, Leipziger Str. 3 – 4, DDR-1086 Berlin; Fernruf: 2236221 oder 2236229; Telex-Nr.: 114420; Bank: Staatsbank der DDR, Berlin, Kto.-Nr. 6836-26-20712.

Chefredakteur: Prof. Dr. Hermann Neels, Zeititzer Weg 25, DDR-7254 Machern; Fernruf: Brandis 3442. Anschrift der Redaktion: Scharnhorststr. 20, DDR-7030 Leipzig; Telefon: 310502.

Veröffentlicht unter der Lizenznummer 1526 des Presseamtes beim Vorsitzenden des Ministerrates der Deutschen Demokratischen Republik.

Gesamtherstellung: VEB Druckerei „Thomas Müntzer", DDR-5820 Bad Langensalza.

Erscheinungsweise: Das Journal „Crystal Research and Technology" erscheint monatlich. Die 12 Hefte eines Jahrganges bilden einen Band. Bezugspreis je Band 480,– DM zuzüglich Versandspesen; Preis je Heft 40,– DM. Der gültige Jahresbezugspreis für die DDR ist der Postzeitungsliste zu entnehmen.

Bestellnummer dieses Heftes: 1082/22/8.

AN (EDV) 35830
02800

Cryst. Res. Technol.	22	1987	8	989–997

Original Papers

H. Kühne, Th. Morgenstern

Akademie der Wissenschaften der DDR, Institut für Halbleiterphysik, Frankfurt/Oder

The Current Understanding of Epitaxial CVD Silicon Layer Doping in the Light of Modelling and Theory Development (V)

Special Features of Doping with Arsenic in the Range of Small Layer Growth Rates

In the SiH_4–HCl–AsH_3–H_2 system of depositing epitaxial silicon doped with arsenic an equilibrium-like doping process has been obtained by way of experiments not only as a limiting case at high deposition temperatures, as had been shown formerly, but also in the range of lower temperatures when the layer growth rate falls below a critical value. By means of introducing the term of a critical rate the theory of dopant incorporation, published in the Parts I–IV of the present report, has been extended to full completeness.

Im SiH_4–HCl–AsH_3–H_2 System tritt bei der Abscheidung von epitaktischen, mit Arsen dotierten Siliciumschichten ein gleichgewichtsnaher Dotandeneinbau nicht nur als Grenzfall für hohe Abscheidungstemperaturen, der schon früher beschrieben wurde, auf, sondern auch im Bereich niedriger Temperaturen, wenn die Schichtwachstumsrate einen kritischen Wert unterschreitet. Die in den Teilen I–IV der vorliegenden Arbeit veröffentlichte Theorie des Dotandeneinbaus wird durch die Einführung des Begriffs „kritische Rate" vervollständigt.

Introduction

In their work on a trapping model of autodoping Wong and coworkers confirmed a peculiarity of the growth rate effect on autodoping phneomena which, in the range of extremely small growth rates, had been observed by Srinivasán. In connection with intended layer doping under steady state conditions Bloem had already formerly obtained a similar effect. In the range below a critical value of the layer growth rate the dopant concentration of the layer decreases with the layer growth rate being reduced. In such experiments the partial pressure of the dopant source in the gas is held at a constant level. On the basis of the commonly accepted model of dopant incorporation at atmospheric pressure conditions (equ. 1) a monotoneous decay of the reciprocal dopant concentration namely proportional to a decreasing growth rate should be expected

$$\frac{p^0_{XH_3}}{(N_{X_{Si}})_{total}} = \frac{1}{A} \cdot (1 + B \cdot v_{epi}) . \quad (1)$$

Contrary to the observations mentioned above, the zero-rate situation should be characterized by the highest doping level possible under the chosen conditions. In the range of high deposition temperatures $B \cdot v_{\text{epi}} \ll 1$ is valid and equ. (1) changes to the form of a solution equilibrium.

Investigating the temperature dependence of growth rate effects on epitaxial silicon layer doping with arsenic, surprisingly, we obtained a lack of growth rate effects on doping too, when in the range of lower deposition temperatures the growth rate had been reduced below a critical value.

The present paper informs about our experimental findings as well as about the extension of the theory of dopant incorporation necessarily been carried out with regard to the real situation of dopant incorporation at small layer growth rates.

Experimental results

Epitaxial CVD layer deposition and doping with arsenic were carried out at atmospheric pressure in excess amounts of hydrogen in a horizontal reactor equipment. Layer deposition was performed using silane-hydrogen chloride mixtures. The partial pressure of arsine was held at a constant level of about $3.5 \cdot 10^{-9}$ atm, the layer growth rate was varied from 0.05 to 1.2 μm $\cdot$ min^{-1}. Different growth rates were adjusted by varying the input pressure of silane as well as that of hydrogen chloride.

Layer thicknesses were obtained from stacking fault measuring, layer doping levels from measuring the sheet resistance of the layer applying a four-point-probe.

In Figure 1 the dependence of dopant incorporation on layer growth rate is shown for four different deposition temperatures. The reciprocal of the arsenic concentration is plotted against the layer growth rate, the temperature is chosen as a parameter.

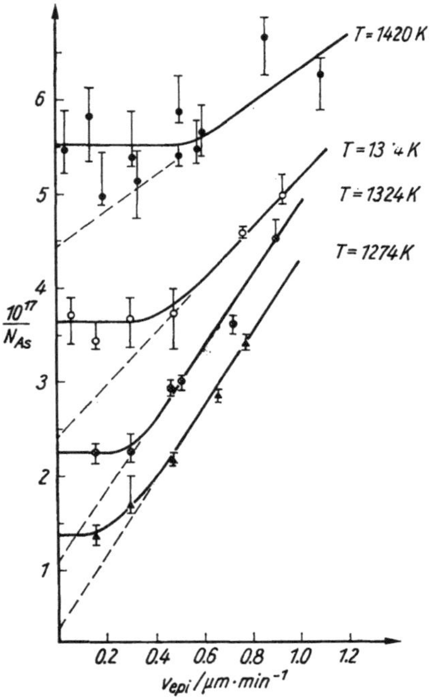

Fig. 1. Reciprocal arsenic concentration as a function of the layer growth rate concerning epitaxial silicon deposition in the SiH_4–HCl–AsH_3–H_2 system, temperature chosen as a parameter. The error bars represent the total deviation from the main value

Contrary to the expectations deduced from equ. (1) both quantities show a linear relationship only in the range of sufficiently large growth rates. Below a critical value of the growth rate (v^*_{epi}) dopant incorporation takes place undisturbed by the fact of layer growth.

At 1274 K the phenomenon has not yet clearly developed. At higher temperatures it more and more becomes evident because v^*_{epi} extends to higher values.

Modification of the theory of dopant incorporation

The theoretical background of the doping process may be characterized by three general steps which will be explained in detail in connection with discussing what we formerly called the "improved" conception of dopant incorporation (cp. Part IV). At first, the decomposition of the dopant source substance by heat has to be taken into consideration. The state of equilibrium is assumed to be reached what in the case of an arsine-hydrogen mixture, is illustrated by equ. (2).

$$AsH_3 \rightleftharpoons AsH_2 \rightleftharpoons AsH \rightleftharpoons As \rightleftharpoons \tfrac{1}{2}\, As_2 \rightleftharpoons \tfrac{1}{4}\, As_4\,. \tag{2}$$

At atmospheric pressure condition the mass balance equation may be drastically simplified since AsH_2-species dominate all the other ones by orders of magnitude.

$$p^0_{AsH_3} = \sum i \cdot p^{eq}_{As_iH_j} \approx p^{eq}_{AsH_2}\,. \tag{3}$$

Secondly, the particles of the decomposition components above all AsH_2-species, are transported to the silicon surface where a certain portion is fixed by adsorption and chemically converted forming arsenic atoms. The latter are consumed in feeding the dopant incorporation as well as in maintaining that reaction step being oppositely directed to the formation reaction. In such a way the dopant incorporation flux (I_{dope}) counterbalances the adsorption and desorption fluxes as shown below:

$$* + AsH_2 \rightarrow As^* + H_2\,, \quad I_{ad} = k_{ad} \cdot (N_0 - N^*) \cdot p^{eq}_{AsH_2} \approx k_{ad} \cdot N_0 \cdot p^{eq}_{AsH_2}, \tag{4}$$

$$As^* + H_2 \rightarrow AsH_2 + *\,, \quad I_{des} = k_{des} \cdot p^{eq}_{H_2} \cdot N_{As^*} \approx k_{des} \cdot p_{total} \cdot N_{As^*}\,, \tag{5}$$

$$I_{dope} = (I_{ad} - I_{des})\,. \tag{6}$$

On the other hand the dopant flux is connected with the dopant concentration of the layer by the growth rate:

$$I_{dope} = (I_{ad} - I_{des}) = v_{epi} \cdot (N_{As_{Si}})_{total} = \frac{v_{epi}}{F_1} \cdot (N_{As_{Si}})_{T_{epi}}\,. \tag{6a}$$

Thirdly, incorporation is completed by the action of a solution equilibrium. It controls the transfer of arsenic atoms from the state of being adsorbed on the surface to that of being located at bulk lattice sites.

$$As^* \rightleftharpoons As_{Si} + * \tag{7a}$$

$$N_{As^*} = \frac{N_0}{K_{inc}} \cdot (N_{As_{Si}})_{T_{epi}} = \frac{F_1 \cdot N_0}{K_{inc}} \cdot (N_{As_{Si}})_{total}\,. \tag{7b}$$

The incorporation equilibrium given with equ. (7) is related to that fraction of incorporated dopants which is in the neutral state at deposition temperature. Together with that other fraction being ionized at deposition temperature, it forms what we call the total amount of arsenic which is identical with the concentration of electrically active arsenic atoms at room temperature. In context to the above concept of an incorporation theory the only possible way of introducing a term like that of a critical layer growth rate is given by equ. (6a), for the growth rate appears in connection with only that equation. The latter defines the dopant flux and its relationship with the total concentration of arsenic atoms in the layer. The dopant flux results from the non-equilibrium situation of the adsorption-desorption process. The critical growth rate, ones reached, defines adsorption-desorption to be in equilibrium. This leads to the following correction of equ. (6a):

$$I_{\text{dope}} = (v_{\text{epi}} - v^*_{\text{epi}}) \cdot (N_{\text{AsSi}})_{\text{total}} = v_{\text{epi}} \cdot \left(1 - \frac{v^*_{\text{epi}}}{v_{\text{epi}}}\right) \cdot (N_{\text{AsSi}})_{\text{total}} = 0\Big|_{v_{\text{epi}} \leqq v^*_{\text{epi}}} . \tag{6b}$$

Characterized by equ. (6b), the dopant incorporation process changes from a kinetically controlled reaction type to a throughout equilibrium situation after the growth rate has been reduced to the value v^*_{epi}, though layer growth is continued.

Correcting equ. (6a) by means of equ. (6b) and taking into account the equs (4) and (5) allow us to formulate the following expression by which steady state arsenic incorporation is described completely.

$$(N_{\text{AsSi}})_{\text{total}} = \frac{K_{\text{inc}} \cdot K_{\text{ad}} \cdot p^{\text{eq}}_{\text{AsH}_3}}{\left[p^{\text{eq}}_{\text{H}_2} + \dfrac{K_{\text{inc}}}{F_1 \cdot N_0 \cdot k_{\text{des}}} \cdot \left(1 - \dfrac{v^*_{\text{epi}}}{v_{\text{epi}}}\right) \cdot v_{\text{epi}}\right]} \equiv$$

$$\equiv \frac{A \cdot p^0_{\text{AsH}_3}}{\left[p_{\text{total}} + B \cdot \left(1 - \dfrac{v^*_{\text{epi}}}{v_{\text{epi}}}\right) \cdot v_{\text{epi}}\right]} = \frac{A \cdot p^0_{\text{AsH}_3}}{\left[1 + B \cdot \left(1 - \dfrac{v^*_{\text{epi}}}{v_{\text{epi}}}\right) \cdot v_{\text{ept}}\right]}\Bigg|_{p_{\text{total}} = 1\,\text{atm}} =$$

$$= A \cdot p^0_{\text{AsH}_3}\,|_{B \cdot v_{\text{epi}} \ll 1,\, v_{\text{epi}} \leqq v^*_{\text{epi}}} . \tag{8}$$

In comparison with the hitherto accepted equ. (1), equ. (8) differs in the additional term $(1 - v^*_{\text{epi}}/v_{\text{epi}})$. This term is causing the kinetics-oriented influence of layer growth on dopant incorporation to disappear not only at high processing temperatures but also at low temperatures when the layer growth rate chosen is equal to or smaller than v^*_{epi}.

From this point of view the general statement given by equ. (9) no longer retains a universal meaning.

$$(N_{\text{XSi}})_{\text{total}} = \frac{1}{v_{\text{epi}}} \cdot I_{\text{XSi}} . \tag{9}$$

It should advantageously replaced by the more fundamental expression:

$$(N_{\text{XSi}})_{\text{total}} = \frac{Q_{\text{XSi}}}{d_{\text{epi}}} . \tag{10}$$

Contrary to equ. (9) the latter does not suggest that layer growth regulates dopant incorporation in any possible case. Equation (9) now is restricted to such doping process versions which are, at least partly, controlled by kinetics. On the other hand equ. (10) excludes the equilibrium-oriented incorporation process from being characterized by using a term like dopant flux.

Discussion of results

Experimental and theoretical results may be discussed on the basis of equ. (8) in specifying the quantities A, B, and v^*_{epi}. In Figure 1 the experimental curves

$(E17/N_{As}) = (f(v_{epi})$ intersect the ordinate at a^* that corresponds to the doping concentration N^*_{As} on one hand and to the growth rate $v_{epi} = v^*_{epi}$ on the other. In addition we define a^0 as the intersection of the extrapolated linear branch of such a curve with the ordinate. It corresponds to both $N_{As} = N^0_{As}$ and $v_{epi} = 0$. Any other point of the curves considered here is defined by a, N_{As}, and v_{epi}.

The slope of such curve as well as the related quantity B of equ. (8) are accessible in the common way by forming the difference quotient. For this it is advantageous to use the following pairs of coordinates: a^0, $v_{epi} = 0$, and a_1, $v_{epi} = 1\ \mu m/min$ as shown in equ. (11)

$$B = \frac{\left(\frac{N^0}{N} - 1\right)}{v_{epi}} = \left(\frac{N^0}{N_1} - 1\right)\Bigg|_{\substack{T_{epi} = \text{const} \\ p^0_{AsH_3} = \text{const}}} = \left(\frac{a}{a^0} - 1\right). \tag{11}$$

In equ. (12) a more general version of equ. (11) is given, to which the quantity a^* is applied:

$$\frac{\left(\frac{N^0}{N_1} - \frac{N^0}{N_2}\right)}{(v_{epi_1} - v_{epi_2})} = \frac{\left(\frac{N^0}{N_1} - \frac{N^0}{N^*}\right)}{(v_{epi_1} - v^*_{epi})} = \frac{\left(\frac{N^0}{N_1} - \frac{N^0}{N^*}\right)}{(1 - v^*_{epi})}\Bigg|_{v_{epi_1} = 1\mu m/min} = \frac{(a - a^*)}{a_0 \cdot (1 - v^*_{epi})}. \tag{12}$$

In combination with equ. (11) the latter allows to get correct values of v^*_{epi}:

$$v^*_{epi} = \frac{(a^* - a^0)}{(a - a^0)}\Bigg|_{a = v_{epi} = 1\mu m/min}. \tag{13}$$

In Figure 2 v^*_{epi} is shown as a function of the reciprocal temperature. The curve possesses a hyperbolic-like shape. In Figure 3 the model-constants A (particle/cm³) and B (S/cm) are shown in the $\log/T^{-1}$ scale. The temperature dependence of A which has been obtained from a^* (cp. Fig. 4) exactly agrees with the value given by RAI-

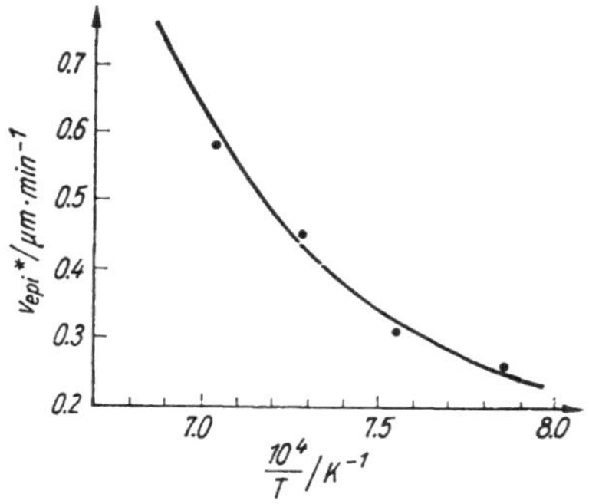

Fig. 2. Dependence of the critical growth rate v^*_{epi} on temperature

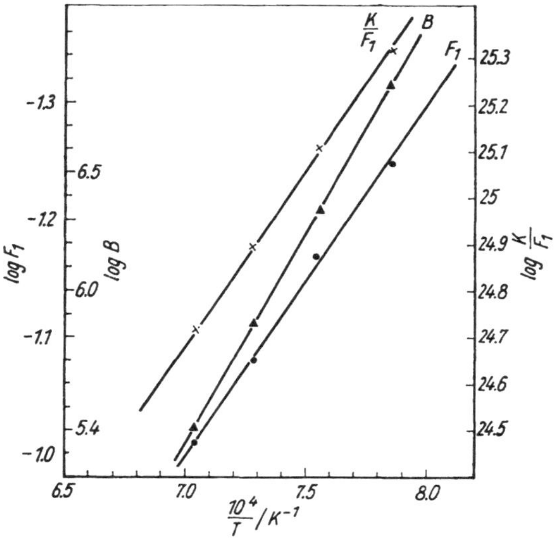

Fig. 3. Temperature dependencies of the model constants A and B as well as of the function F_1

CHOUDHURY and SALKOVITZ, i.e. -33.3 kcal/mole and -32 kcal/mole, respectively. It also agrees with that obtained by REIF et al. (-30 kcal/mole) though the latter value corresponds to a layer agrowth rate of $v_{epi} = 1.2$ µm/min. Contrary to REIF we obtained a temperature dependence as small as -8 kcal/mole at $v_{epi} = 1$ µm/min. The different results may possibly be explained by the absence of chlorine in the system used by REIF and coworkers. In the above context it should be of interest that a_0 possesses a temperature dependence of -59.5 kcal/mole, which is in good agreement with the value formerly derived by KÜHNE et al. for the As–H–Si system (-60 kcal/mole). The dependencies of a, a^0, and a^* on temperature are shown in Figure 4.

Figure 3 also shows how F_1 depends on temperature (KÜHNE).

$$F_1 = [1 + 1/2 \cdot \exp\,(E_{dope} - E_F)/kT]^{-1} = \frac{(N_{X_{Si}})_{T_{epi}}}{(N_{X_{Si}})_{total}}. \tag{14}$$

Assuming the calculated values of F_1 to be correct enables us to obtain additional informations regarding the theoretical quantities of the incorporation law. Yet, in this context the more straightforward oriented "consequent" conception of dopant incorporation (cp. Part IV) should advantageously be used instead of the "improved" conception, what might be demonstrated by comparing the correspondent incorporation law, given below, with equ. (8).

$$(N_{As_{Si}})_{total} = \frac{K \cdot p^{eq}_{AsH_3}}{F_1 \cdot \left[p^{eq}_{H_2} + \frac{v_{epi}}{F_1 \cdot k_2} \cdot \left(1 - \frac{v^*_{epi}}{v_{epi}}\right)\right]} =$$

$$= \frac{A \cdot p^0_{AsH_3}}{\left[p_{total} + B \cdot \left(1 - \frac{v^*_{epi}}{v_{epi}}\right) \cdot v_{epi}\right]} = \frac{A \cdot p^0_{AsH_3}}{\left[1 + B \cdot \left(1 - \frac{v^*_{epi}}{v_{epi}}\right) \cdot v_{epi}\right]}\Bigg|_{p_{total} = 1\,atm}$$

$$= A \cdot p^0_{AsH_3}\,|_{B \cdot v_{epi} \ll 1,\; v_{epi} \leq v^*_{epi}}\,. \tag{15}$$

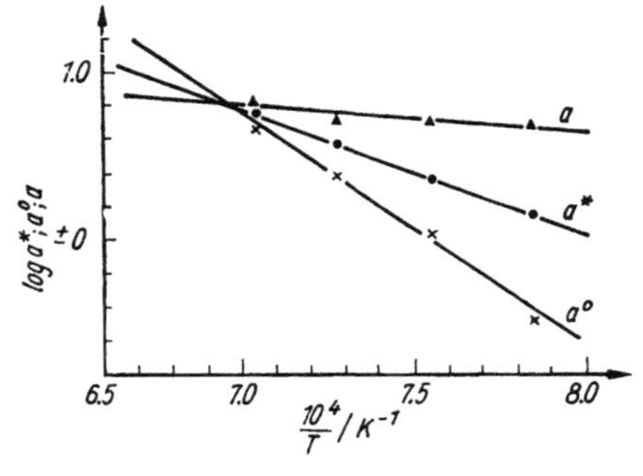

Fig. 4. Arrhenius plot of the experimentally related quantities a^* ($v_{epi} = v^*_{epi}$), a_0 ($v_{epi} = 0$), and a_1 ($v_{epi} = 1$ µm/min)

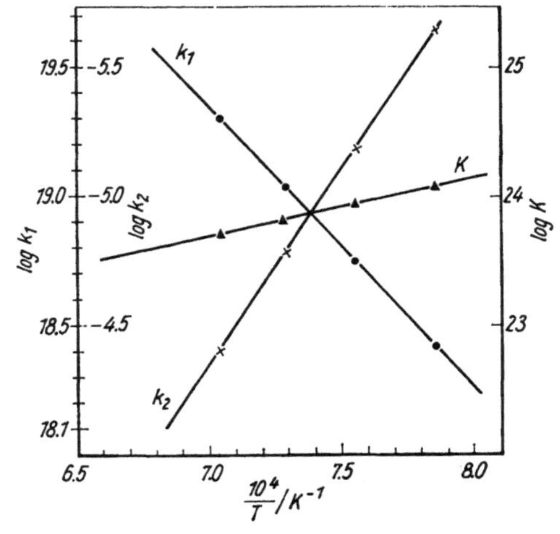

Fig. 5. Arrhenius plot of the theory-related quantities K, k_1, and k_2

Table 1

Quantity	reaction step	temperature dependence/ kcal/mole	enthalpy kcal/mole
a^*; ($v_{epi} \leqq v^*_{epi}$)		-33.3	
a^0; ($v_{epi} = 0$)		-59.5	
a_1; ($v_{epi} = 1\ \mu m \cdot min^{-1}$)		-7.9	
$A = K/F_1$ ($v^*_{epi} \geqq v_{epi} > 0$)	$AsH_2 \rightleftharpoons As_{Si} + H_2$	$+33.3$	
$F_1 = \frac{(N_{As_{Si}})_{T_{epi}}}{(N_{As_{Si}})_{total}}$		-13.9	$\Delta H_{n_1} = +29.5$
$\frac{d \ln F_1}{d \frac{1}{RT}}$ $\frac{d \ln n_1}{d \frac{1}{RT}}$			
$K = A \cdot F_1$		$+19.4$	$\Delta H_K = -3.8$
$k_1 = A/B$	$AsH_2 \rightarrow As_{Si} + H_2$	-47.9	
$k_2 = k_1/K$	$As_{Si} + H_2 \rightarrow AsH_2$	-67.3	

Taking equ. (15) as a base we obtain K from A, k_1 from A/B, and k_2 from k_1/K. Results of such calculations are shown in Figure 5. The temperature dependencies of those quantities are summarized in Table 1.

In the case of K it is not such a simple step as in general use, to conclude from the value of its dependence on temperature to that of its enthalpy. This peculiar circumstance is caused by the fact that the temperature dependence of F_1 is determined by the energy gap of silicon. The latter, as has been proved by THURMOND, is a quantity related to Gibbs' energy rather than to an enthalpy term. In consequence we find that the enthalpy of incorporation ($\Delta H_K = -3.8$ kcal/mole) is characterized by a negligibly small value (cp. equ. (16)) in contrast to the temperature dependence of incorporation (33.3 kcal/mole).

$$\Delta H_K = -\frac{d \ln A}{d \frac{1}{RT}} + \Delta H_{F_1} \approx -\frac{d \ln A}{d \frac{1}{RT}} + \Delta H_{n_1} = $$
$$= -33.3 + 29.5 = -3.8 \text{ kcal/mole} . \tag{16}$$

This particular circumstance effectuates the quantity of the overall enthalpy of arsenic incorporation to be nearly entirely determined by the ionization equilibrium of the arsenic atoms in the silicon layer (cp. Part II).

Figure 5 also shows that k_2 is stronger influenced by temperature than k_1, and both are influenced in such way that the corresponding reaction steps are accelerated by increasing temperature. As to layer growth influence on arsenic incorporation, Figure 6 shows the temperature dependence of that effect in the version of the uncorrected model (equ. (1)). Accordingly an equilibrium-like incorporation process should exist only as a limiting case at sufficiently high deposition temperatures. The corrected version (equ. (8) or (15)) leads to results shown in Figure 7. They differ from the uncorrected ones in that the curves intersect the zero-level of the ordinate

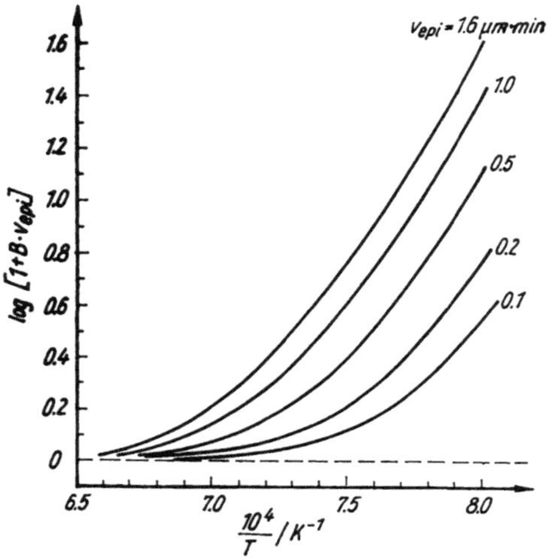

Fig. 6. The uncorrected layer growth rate effect on arsenic incorporation with respect to equ. (1): $(1 + B \cdot v_{epi})$ as a function of temperature, layer growth rate chosen as a parameter

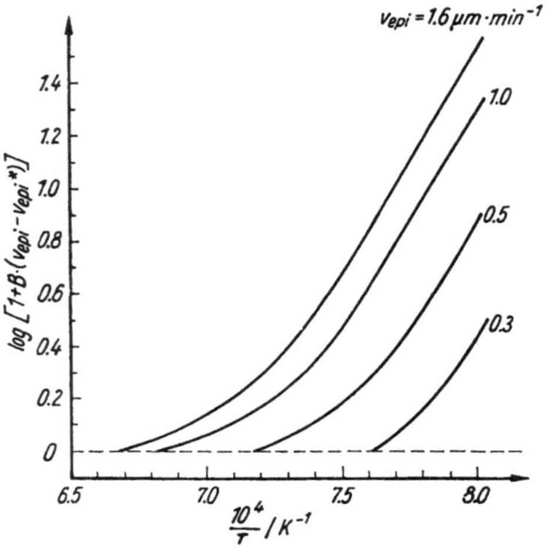

Fig. 7. The corrected layer growth rate effect on arsenic incorporation with respect to equ. (8): $[1 + B \cdot (1 - v^*_{epi}/v_{epi}) \cdot v_{epi}$ as a function of temperature, layer growth rate chosen as a parameter

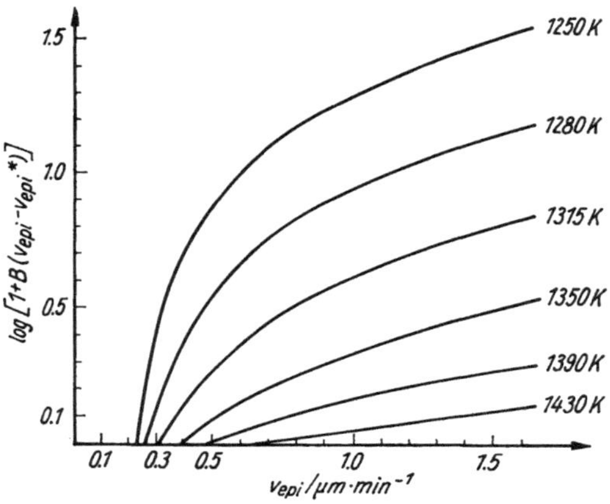

Fig. 8. The correct layer growth rate effect on arsenic incorporation with respect to equ. (8): $[1 + B \cdot (1 - v^*_{epi}/v_{epi}). v_{epi}$ as a function of the layer growth rate, temperature chosen as a parameter

at that temperature at which the v_{epi}-value is identical with v^*_{epi} (cp. Fig. 2). Therefore in Figure 7 the zero-level is not asymptotically approached as in Figure 6.

In Figure 8 the dependence of the layer growth effect on arsenic incorporation is shown as a function of the layer growth rate. At high deposition temperatures the growth rate effect is small, the curves decrease in a nearly linear way until v^*_{epi} is reached and the growth rate effect vanishes. At low deposition temperatures the growth rate influence on doping strongly decreases when v_{epi} approaches v^*_{epi}.

From Figure 8 it is understandable that HERRING's experimental result of silicon doping with arsenic under low pressure conditions can be interpreted on the basis of a pure equilibrium incorporation mechanism ($T_{epi} = 1353$ K, $v_{epi} = 0.3 - 0.35$ µm/per min) when a two-different-particle-incorporation is taken into account (KÜHNE, KOKOVIN et al.).

Conclusions

The theory of dopant incorporation derived in the previously published Parts I—IV has been extended to full completeness by introducing a term called critical layer growth rate. It regulates dopant incorporation to be governed by an equilibrium-like mechanism or, at least partly, controlled by kinetical reaction steps according as v_{epi} is chosen smaller or larger than v^*_{epi}. In consequence, dopant flux is a quantity that should be used with respect to characterizing kinetically influenced doping processes whereas in the case of an equilibrium-like process the more universal term of dopant dose should be used because dopant flux then is zero by definition.

References

BLOEM, J.: J. Cryst. Growth **13/14**, 302 (1979)
HERRING, R. B.: Electrochem. Soc. Fall Meeting (1979), Los Angeles
KOKOVIN, G. A. TESTOVA, A. A., TITOV, A. A., MORGENSTERN, Th., KÜHNE, H.: Cryst. Res. Technol. **20**, 1583 (1985)
KÜHNE, H. (parts I—IV): ibid. **22**, 487, **669**, **767**, **883** (1987)
KÜHNE, H.: ibid. **17**, 1217, 1379 (1982)
KÜHNE, H.: ibid. **17**, 181 (1982)
KÜHNE, H., SPERLING, R.: Kristall u. Technik **15**, 403 (1980)
RAI-CHOUDHURY, P., SALKOVITZ, E. J.: J. Cryst. Growth **7**, 353 (1970)
REIF, R., KAMINS, T. J., SARASWAT, K.: J. Electrochem. Soc. **125**, 1860 (1978)
SRINIVASAN, G. R.: ibid. **125**, 146 (1978)
THURMOND, C. D.: ibid **122**, 1133 (1975)
WONG, M., REIF, R.: IEE Trans. Electr. Dev. **ED-22**, 83 (1985)
WONG, M., REIF, R., SRINIVASAN, G. R.: ibid. **ED-22**, 89 (1985)

(Received December 5, 1986)

Authors' address:

Dr. H. KÜHNE, Dr. Th. MORGENSTERN
Institut für Halbleiterphysik der Akademie der Wissenschaften der DDR
W.-Korsing-Str. 2, PSF 409
Frankfurt/Oder
DDR-1200

Cryst. Res. Technol. | 22 | 1987 | 8 | 999–1007

S. Schwetlick, W. Seifert, E. Butter, W. Hörig, R. Pickenhain, R. Schwabe

Karl-Marx-Universität Leipzig, Sektionen Chemie und Physik,
Arbeitsgemeinschaft „AIIIBV-Halbleiter"

Incorporation of Nitrogen into Galliumarsenide Grown by Chloride VPE

The incorporation of nitrogen into GaAs in the halide-VPE system using NH_3 as the doping source is investigated. The concentration of incorporated N depends on the NH_3 flow rate and increases with decreasing deposition temperature. Concentrations up to 6×10^{17} cm^{-3} could be detected far above the calculated equilibrium value.

By a comparison of SIMS-data with those obtained from IR it is found that most of the N-atoms are situated substitutionally on As lattice sites.

Normal-pressure 2 K photoluminescence measurements show N-related luminescence peaks at 1.508 and 1.496 eV. Most probably this luminescence arises from complexes of N with undeliberately introduced impurities. For the 1.508 eV center a complex Si_{Ga}–N_{As} is discussed. — A new N-related metastable deep center with the energy levels E_C–0.33, E_C–0.41, and E_C–0.68 eV could be found by DLTS.

Der Einbau von Stickstoff in GaAs im Halogentransport-VPE-System mit NH_3 als Dotierquelle wird untersucht. Die Konzentration des eingebauten Stickstoffs hängt ab von der NH_3-Flußrate und wächst mit fallender Abscheidungstemperatur. Konzentrationen bis zu 6×10^{17} cm^{-3} konnten bestimmt werden — weit oberhalb der berechneten Gleichgewichtskonzentration. — Durch einen Vergleich von SIMS- und IR-Daten wird gefunden, daß die meisten der N-Atome substitutionell auf As-Plätzen eingebaut werden.

Messungen der 2K-Photolumineszenz unter Normaldruck ergeben mit Stickstoff in Verbindung stehende Lumineszenzpeaks bei 1.508 und 1.496 eV. Diese Lumineszenz rührt höchstwahrscheinlich her von Komplexen des N mit unbeabsichtigt eingebauten Verunreinigungen. Für das 1.508 eV-Zentrum wird ein Komplex Si_{Ga}–N_{As} diskutiert. — Durch DLTS konnte ein neues tiefes Zentrum mit den Energiezuständen E_c–0.33, E_c–0.41 und E_c–0.68 eV gefunden werden.

1. Introduction

The incorporation of the isoelectronic impurity nitrogen into GaP and GaP_xAs_{1-x} is widely used to enhance the recombination efficiency of optoelectronic devices.

Using NH_3 as the doping source a second effect could be observed: the presence of NH_3 in the vapour phase reduces the concentration of free carriers in chloride VPE grown GaP (Stringfellow et al. 1975a), GaAs (Stringfellow, Hom) as well as InP (Sun, Wessels).

Stringfellow and Hom attributed this effect to the gettering of residual silicon by ammonia in the gas phase during crystal growth:

$$3\, SiCl_nH_{4-n}(g) + 4\, NH_3(g) \rightarrow Si_3N_4(s) + 3n\, HCl(g) + (12 - 3n)\, H_2(g)\,.$$

An effective incorporation of N into GaAs was discounted since a thermodynamic calculation indicated that the solubility of nitrogen in GaAs should be less than 10^{14} cm^{-3}.

WOLFORD et al., however, found a direct evidence of nitrogen incorporation through an observation of N-related lines appearing at pressures above 2.4 GPa in the low-temperature photoluminescence of intentionally N-doped GaAs. A similar observation made LEROUX et al. in GaAs samples grown by MBE using BN-crucibles. About N-related peaks appearing even in normal pressure low temperature photoluminescence spectra reported SCHWABE et al. The concentration of the N-related centers found in GaAs grown in the presence of NH_3 in the halide VPE system could be estimated to exceed 5×10^{16} cm^{-3}. This result has been confirmed by infrared optical investigations (RIEDE et al.). Finally, LUPAL et al. determined N-concentrations in the order of 10^{17} cm^{-3} in halide VPE grown GaAs by SIMS.

From these experiments follows that nitrogen is incorporated into GaAs far above the calculated equilibrium solubility. Since kinetical effects seem to be responsible for this behaviour our intention is to study some relations between N-incorporation and growth mechanism of GaAs. Furthermore, we discuss some aspects concerning the defect structure of the centers related to N in GaAs.

2. Experimental

The growth experiments were performed in the Ga–GaAs(crust)–$AsCl_3$–H_2–NH_3 system varying the NH_3 inlet pressure and the deposition temperature. The difference between source and deposition zone temperature was kept constant at 40 °C. Typical growth conditions are given in Table 1.

Table 1
Growth parameters of GaAs:N epitaxial layers

No	T_{depos} K	f^0_{total} sccm	$p^0_{AsCl_3}$ kPa	$p^0_{NH_3}$ kPa	thickness μm
533	1013	600	0.44	0	19
532	1013	600	0.44	0.8	21
530	1013	600	0.44	1.6	16
531	1013	600	0.44	3.3	17
534	1013	600	0.44	4.2	19
554	1043	600	0.44	3.3	105
557	1028	600	0.44	3.3	108
556	1013	600	0.44	3.3	102
555	983	600	0.44	3.3	32

The substrates used were SI–GaAs:Cr and n-GaAs:Te (100)-oriented 2° off toward [110]. They were treated with a conventional trichloroethylene/acetone/methanol/H_2SO_4, H_2O_2, H_2O/H_2O/methanol/isopropanol cleaning procedure and in-situ etched prior to growth. The samples were analyzed by PL-, IR-, SIMS-, and DLTS measurements. The photoluminescence spectra were obtained by exciting the samples with an Ar^+-laser ($\lambda = 515.4$ nm) at 2 K. The pump level was 0.5 W cm^{-2}, the spectra resolution amounted to about 0.04 meV. For the infrared optical measurements the substrates were removed from the samples. The transmittance and reflectance of the parallel-sided substrate-free epitaxial layers were measured at 300 K in the wave number range $\bar{\nu} = 400$ to 600 cm^{-1} with unpolarized radiation using a Perkin-Elmer spectrophotometer model 580 B. For SIMS measurements a 14.5 eV Cs^+-primary ion excitation was used. The inaccuracy of the measured N-concentration was estimated to be about 20%. DLTS-spectra were measured between 80 and 480 K using a conventional lock-in DLTS equipment with a sensitivity of $\Delta C/C_0 = 10^{-6}$.

3. Results

In Figure 1 is shown the temperature dependence of the growth rate in the presence of various amounts of ammonia. The growth rate is nearly independent of the NH_3 input pressure. It decreases with decreasing deposition temperature over the whole temperature range. That means for an exothermic deposition reaction that the growth process must be limited by a kinetically controlled reaction step.

In Table 2 electrical properties of GaAs grown with different partial pressures of NH_3 and at different deposition temperatures are given. It can be seen that the addition of NH_3 to the ambient decreases the carrier concentration. This carrier-reducing effect of NH_3 is the more pronounced the higher the deposition temperature resulting in SI-GaAs at 755 °C and even in p-material at 770 °C. Details concerning the electrical properties will be discussed in a forthcoming paper.

Photoluminescence spectra measured at 2 K are shown in Figures 2 and 3. The admixture of NH_3 to the growth ambient has two effects: firstly, we observe a significant reduction of the half width of the near band gap luminescence bands (1.50 — to 1.52 eV) attributed to the radiative decay of excitons bound to impurities. This effect is partly caused by the reduction in free carrier concentration due to the NH_3. Secondly, we observe new transitions centered at 1.508 eV (NX_1) and 1.496 eV (NX_2) with a distinct fine structure (tripletts). Their intensities relative to the near band gap luminescence bands depend on the ammonia flux (Fig. 2). Furthermore, their intensities decrease with increasing deposition temperature (Fig. 3). Similar luminescence bands were found in GaAs:N grown by the Ga/HCl/AsH_3 deposition technique (SCHWABE et al.). Undoubtedly, these peaks are caused by the radiative recombination of excitons bound to N-centers incorporated into the GaAs.

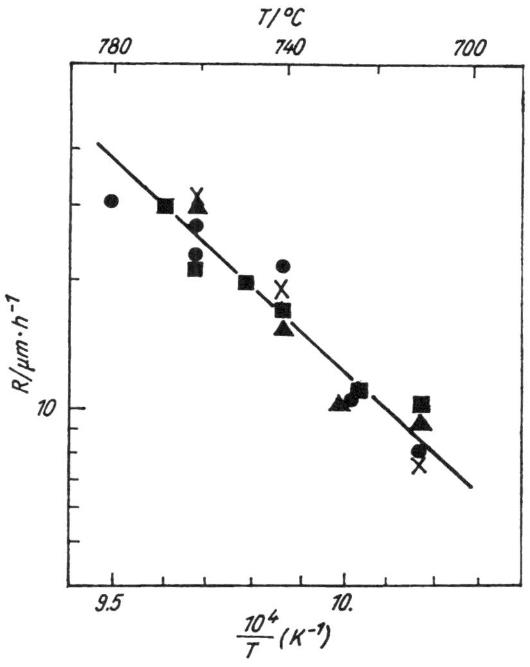

Fig. 1. Growth rate of GaAs: N in dependence on temperature. $p^0_{NH_3}$/kPa: 0 (×), 0.83 (●), 1.66 (▲), 3.33 (■)

Table 2

Electrical properties of GaAs:N eiptaxial layers

No	n, p/cm^{-3}	μ/cm^2/Vs
533	$4.2 \cdot 10^{15}$	5200
532	$4.5 \cdot 10^{14}$	6300
530	$2.2 \cdot 10^{14}$	5200
531	$1.4 \cdot 10^{14}$	5500
534	$7.1 \cdot 10^{13}$	6400
554	$1.3 \cdot 10^{11}$ (p)	280
557	semi-insulating	
556	$1.2 \cdot 10^{15}$	3000
555	$3.1 \cdot 10^{15}$	4600

In Table 3 the N-concentrations measured by SIMS and estimated by IR-absorption intensities of the local mode for samples grown at various substrate temperatures are given.

The SIMS measurements include a thickness of about 5 µm of the layer and show a constant value for the N concentration. The IR values are calculated from the absorption intensities of the 58.2 meV local mode assuming that the absorption coefficient of N_{As} in GaAs has the same value as N_P in GaP (RIEDE et al.). Obviously, the results obtained by both methods are adequate. It is found that in analogy to the observed luminescence intensities of NX_1 and NX_2 in PL the N concentrations increase with decreasing deposition temperature. However, the absolute concentrations are different. An estimation of the concentration of the N-related center NX_1 in analogy to SCHWABE et al. results in $10^{15}-10^{16}$ cm^{-3} in samples showing intense NX_1 luminescence, whereas the "chemical" concentration measured by SIMS exceeds the range of 10^{17} cm^{-3}. This difference may be due to different species incorporated into the GaAs lattice as will be discussed in the next section.

A further information about N incorporation we obtained by deep level transient spectroscopy (DLTS). In addition to the EL 2 center observed in all samples the samples doped with N show a new metastable center. The concentration of this center depends in the same manner on growth conditions as the N-concentration, i.e., it increases with increasing NH_3 flux, and it decreases with increasing growth temperature. In its absolute concentration it reaches the net doping concentration. The

Table 3

Impurity concentrations in GaAs:N epitaxial layers

No	T_{depos}/K	$p^0_{NH_3}$/kPa	[N]/cm^{-3}		[Si]/cm^{-3}
			IR	SIMS	SIMS
533	1013	0	not determined		$1 \cdot 10^{16}$
554	1043	3.3	$1.5 \cdot 10^{17}$	$1.7 \cdot 10^{17}$	$\leqq 4 \cdot 10^{15}$*)
557	1028	3.3	$4.5 \cdot 10^{17}$	$2.0 \cdot 10^{17}$	$\leqq 4 \cdot 10^{15}$*)
556	1013	3.3	$6.0 \cdot 10^{17}$	$5.0 \cdot 10^{17}$	$\leqq 4 \cdot 10^{15}$*)
555	983	3.3	not deter.	$4.0 \cdot 10^{17}$	$\leqq 4 \cdot 10^{15}$*)

*) detection limit for Si

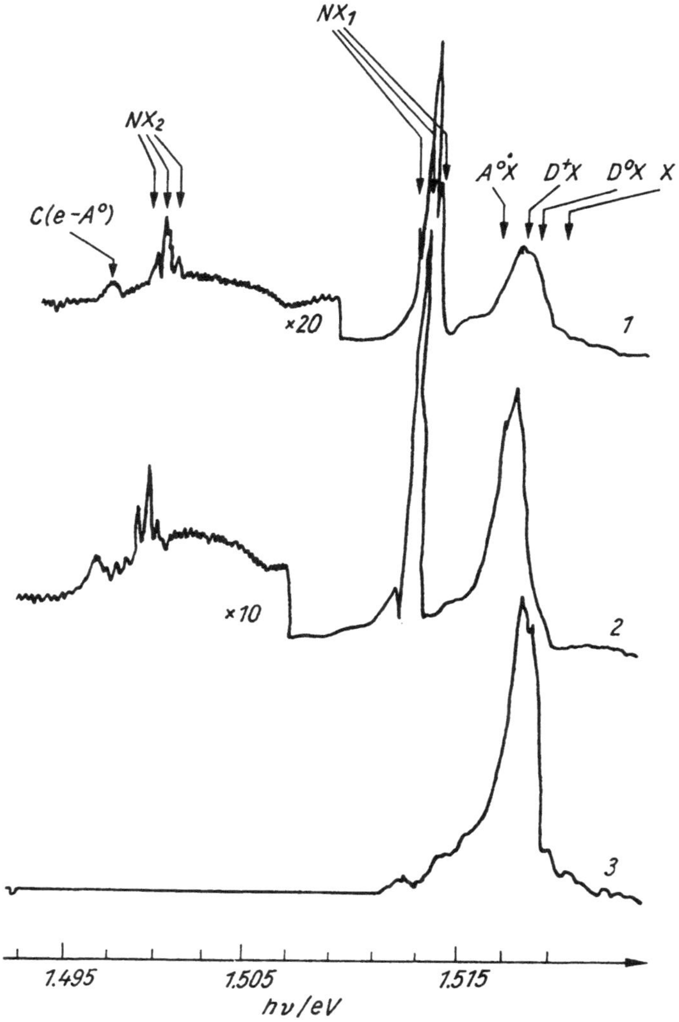

Fig. 2. Photoluminescence (2 K) of GaAs:N samples grown at different deposition temperatures. $p^0_{NH_3} = 3.33$ kPa; $T_{substr.}$ °C: 710 (1), 740 (2) 780 (3)

thermal activation energies of the different states of this center are $E_C - 0.33$ eV, $E_C - 0.41$ eV, and $E_C - 0.68$ eV. More details will be published in a forthcoming paper (PICKENHAIN et al.).

Obviously, the main result is that there exist different N-related centers in GaAs grown in the presence of NH_3. The concentration of N substituting As lattice sites, estimated by IR measurements, agrees well with the "chemical" concentration, measured by SIMS. It is found in the order as high as mid 10^{17} cm^{-3}.

The optically observable center NX_1, seen in PL, seems to be present in concentrations of $<10^{16}$ cm^{-3}. There exists also a further center NX_2 the concentration of which could not be estimated.

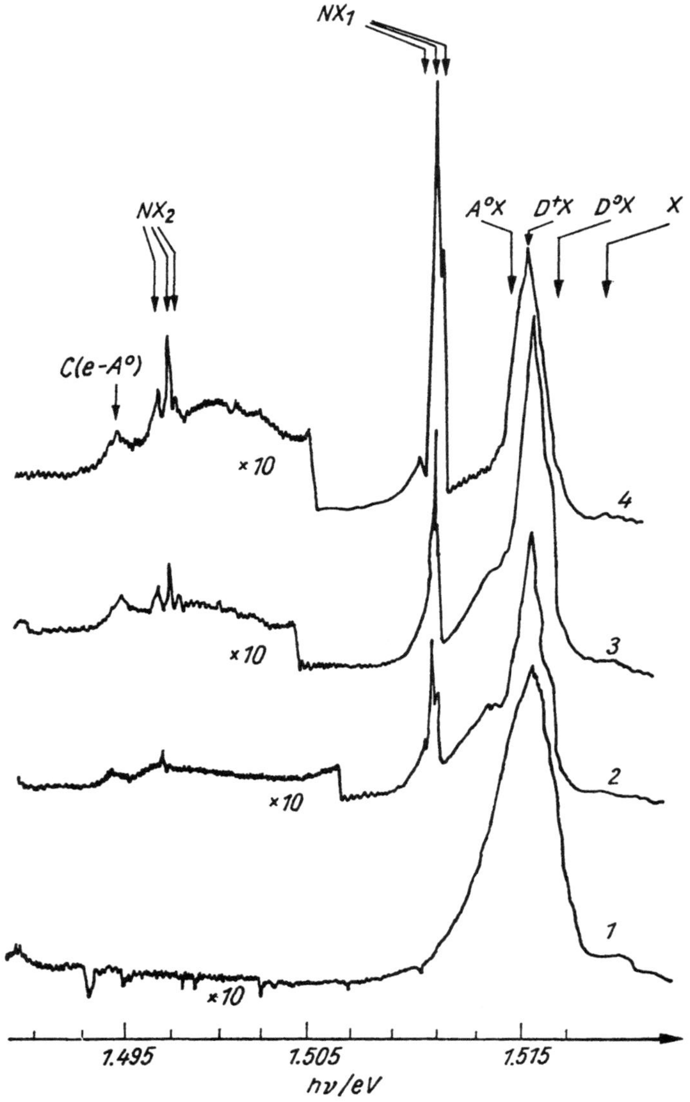

Fig. 3. Photoluminescence (2 K) of GaAs:N samples grown with different partial pressures of NH_3. $T_{substr.}$ = 740 °C, T_{source} = 780 °C. $p^0_{NH_3}$/kPa: 0 (1), 0.83 (2), 1.66 (3), 3.33 (4)

Finally, DLTS measurements show that also N-related deep levels are produced with concentrations reaching the net doping level, i.e. in the order of 10^{14} cm^{-3}. But this is only a raw value since the net doping level is the upper detection limit in DLTS.

The concentrations of all these centers depend on the NH_3 flux during growth and increase with decreasing deposition temperature.

4. Discussion and Conclusions

The experimental results show that the presence of NH_3 during growth induces several nitrogen related defect centers in the GaAs crystal which are different in their physical properties and concentrations.

The total concentration of incorporated nitrogen is some magnitudes higher than the expected equilibrium values calculated by STRINGFELLOW and HOM in an earlier paper. Even the observed temperature dependence stands in opposition to the thermodynamically predicted increase of N incorporation with increasing deposition temperature which follows from equation

$$\frac{\partial \ln x^{s}_{GaN}}{\partial (1/T)} = -\frac{(\Delta_R H^{\theta}_{GaN} + \Omega^{s})}{R} \approx \frac{-(-70 + 190)\ \text{kJ/mole}}{R} < 0$$

with

x^{s}_{GaN} mole fraction of solid GaN in GaAs

$\Delta_R H^{\theta}_{GaN}$ reaction enthalpy (taken as a constant in the temperature range considered here)

Ω^{s} interaction parameter GaN/GaAs, calculated by the delta lattice parameter model (DLP)

Hence, this simple thermodynamic consideration does not reflect the experiment.

A similar behaviour is known from GaP:N too. It could be shown that the relation $[N]_{exp.}/[N]_{calc.}$ becomes larger than one as soon as the growth of the GaP epitaxial layers changed from the mass transport controlled to the surface-kinetically controlled deposition mechanism (STRINGFELLOW et al. 1975b). Following, the observed enhanced N-incorporation seems to be a consequence of the dominant kinetically controlled growth process in our deposition system. The rate-limiting step in this regime is thought to be the reactive desorption of chlorine by molecular hydrogen (THEETEN et al.):

$$As_4(g) \rightleftharpoons 2\,As_2(g) \rightleftharpoons 4\,As\,(surf) \quad (1)$$

$$GaCl(g) \rightleftharpoons GaCl\,(surf) \quad (2)$$

$$GaCl\,(surf) + As\,(surf) \rightleftharpoons As\text{–}Ga\text{–}Cl\,(surf) \quad (3)$$

$$As\text{–}Ga\text{–}Cl\,(surf) + H_2(g) \rightleftharpoons [As\text{–}Ga\text{–}Cl\text{–}H\text{–}H]^*\,(surf) \quad (4)$$

$$[As\text{–}Ga\text{–}Cl\text{–}H\text{–}H]^*\,(surf) \rightarrow GaAs(s) + HCl(g) + H\,(surf) \quad (5)$$

$$2\,H\,(surf) \rightleftharpoons H_2(g)\,. \quad (6)$$

Obviously, in the presence of ammonia also the following reactions have to be taken into consideration:

$$NH_3(g) \rightleftharpoons H_{3-x}N\,(surf) + x\,H(surf)\,, \quad (7)$$

$$GaCl\,(surf) + H_{3-x}N\,(surf) \rightleftharpoons H_{3-x}N\text{–}Ga\text{–}Cl\,(surf)\,, \quad (8)$$

$$H_{3-x}N\text{–}Ga\text{–}Cl\,(surf) + H_2(g) \rightleftharpoons [H_{3-x}N\text{–}Ga\text{–}Cl\text{–}H\text{–}H]^*\,(surf)\,, \quad (9)$$

$$[H_{3-x}N\text{–}Ga\text{–}Cl\text{–}H\text{–}H]^*\,(surf) \rightarrow GaN\,(in\ GaAs) + HCl\,(g) + (4 - x)\,H\,(surf)\,. \quad (10)$$

Since steps (5) and (10), resp., are the kinetic barriers the steps before should be in equilibrium, i.e. the relative surface concentrations of As–Ga–Cl- and $H_{3-x}N$–Ga–Cl-complexes in kink positions are determined by an adsorption-desorption equilibrium. For our growth conditions (with $p^0_{NH_3} > p^0_{As\,total}$) it is reasonable to assume an N/As-ratio on the surface far above the equilibrium ratio in the solid. That means, in step

(10) a large excess of GaN will be produced. For the case that the subsequent decomposition reaction

$$GaN \text{ (in GaAs)} \rightarrow Ga_{Ga} + V_{As} + 1/2\, N_2(g) \tag{11}$$

is kinetically hindered, the N will be "trapped" as soon as the reaction rate of step (11) becomes slower than the growth of the GaAs crystal itself. (Concerning the "trapping mechanism" see GILING and DE MOOR). The decomposition reaction (11) is a comparatively slow process, its slowness is of crucial importance in the vapour deposition of GaN. We found, for instance, that pure GaN epitaxy layers decompose under vacuum with a rate of about 10^{-2} µm/min at 950 °C and 10^{-3} µm/min at 850 °C (SEIFERT, KOKOVIN). Even bearing in mind that somewhat other growth conditions determine the growth of GaAs:N (lower deposition temperature, H_2 ambient, GaAs surface) we assume that it is the slowness of reaction (11) causing the trapping of N into the solid GaAs. This mechanism seems to be the best explanation for the observed nonequilibrium incorporation of N into GaAs under surface-kinetically controlled growth conditions.

Concerning the nature of the N-related centers from the good agreement between IR- and SIMS-data we conclude that most of the nitrogen incorporated into the GaAs crystal is situated on arsenic lattice sites surrounded by four Ga-atoms. The observed localized mode at $469.7 \pm 0.3\ cm^{-1}$ (58.2 meV) (RIEDE et al.) is found to be in excellent agreement with the theoretical estimate of $\bar{\nu}_{loc} = 469\ cm^{-1}$ for the frequency of a nitrogen atom substituting As in GaAs.

However, in contrast to the indirect semiconductor GaP, this isolated nitrogen in GaAs doesn't create any energy levels within the band gap. WOLFORD et al. were able to detect nitrogen bound exciton luminescence only applying high hydrostatic pressure. The luminescence peaks NX_1 and NX_2 in our samples have their origin in other centers containing the N atom as one constituent. The triplet structure of NX_1 and NX_2 hints to a surrounding with a certain deviation from tetrahedral symmetry. We postulate that there is a complex formation between the N and other impurity atoms. The main impurity in our hot-wall growth system is Si causing the n-type conductivity in our N-free GaAs with electron concentrations in the range of about $10^{16}\ cm^{-3}$. This is also the Si doping level determined by SIMS (see Tab. 3).

A complex formation between N_{As} and Si_{Ga} results in a center with C_{3v} symmetry in accordance with the found Zeeman pattern of NX_1 reported by SCHWABE et al. Moreover the formation of such complexes seems to be also supported by the fact that the Si–N bond will be already performed in the vapour phase since in the presence of NH_3 most of the Si species should be gettered in the form of $(Si_xN_yH_z)_n$-polymers and $Si(NH_2)_2$, respectively.

Clearly, in this complex the frequency of the local mode would be lowered due to the lower mass of Si instead of Ga. Using the model of KACHARE for calculation of the local modes of N surrounded by 4 Ga atoms and N surrounded by 4 Si atoms one calculates the frequency of the local mode of N in a N_{As}–Si_{Ga}-complex with the simplification

$$h\nu_{loc}(SiGa_3\text{–}N) = \frac{h\nu_{loc}(Si_4\text{–}N) + 3h\nu_{loc}(Ga_4\text{–}N)}{4}$$

to be $h\nu_{loc} = 54.7$ meV. The existence of a phonon replica of NX_1 at $h\nu = 1.4515$ eV with an 56.7 meV difference to NX_1 could be found in the PL spectra.

In conclusion, we have demonstrated that the incorporation of nitrogen into GaAs during VPE in the presence of NH_3 results in nitrogen concentrations far above previously calculated equilibrium values. This behaviour could be explained by the kinetics of the surface reaction. In a surface-reaction controlled growth regime, the

NH_x–Ga–Cl- and the As–Ga–Cl complexes are competing surface species, the relative concentration of which is determined by an adsorption-desorption equilibrium. Contrary, the incorporation step is a non-equilibrium process. At usual growth rates the equilibration between N_{solid}/N_{gas} is slower than the growth process itself. Thus, it is expected that the relatively high concentration of NH_x–Ga–Cl complexes leads to a high supersaturation of N at the surface with a subsequent trapping of N into the solid.

By a comparison of SIMS data with those obtained from IR measurements it is followed that most of the nitrogen atoms are incorporated substitutionally on As lattice sites. Photoluminescence measurements have shown that part of nitrogen atoms form complexes with impurity atoms, the most probable of which is Si. Thus the N-related luminescence at 1.508 eV, reported in a previous paper, seems to originate from Si_{Ga}–N_{As} complexes.

Further N-related centers, found in PL spectra as well as by DLTS exist, but in concentrations some orders of magnitude below the total concentration of nitrogen.

We wish to thank Prof. K. UNGER for his support and interest, Dr. H. NEUMANN for helpful discussions and TSCHAN UK SONG for cowork in an early stage of this investigations.

References

GILING, L. J., DE MOOR, H. H. C.: Proceedings of the 4th European Conference on CVD, Eindhoven, ed. J. BLOEM, G. VERSPUI and L. R. WOLFF, 1983, p. 184

KACHARE, A. H., SPITZER, W. G., KAHAN, A., EULER, F. K., WHATLEY, T. A.: J. Appl. Physics. 44, 4393 (1973)

LUPAL, M. W., KLOTH, B., PICHTIN, A. N., RICHTER, K., TRAPP, M.: Neorg. Mater. 22, 186 (1986)

LEROUX, M., NEU, G., CONTOUR, J. P., MASSIES, J., VERIE, C.: J. Appl. Physics 59, 2996 (1986)

RIEDE, V., NEUMANN, H., SOBOTTA, H., SCHWABE, R., SEIFERT, W., SCHWETLICK, S.: phys. stat. sol. (a) 93, K151 (1986)

SCHWABE, R., SEIFERT, W., BINDEMANN, R., AGEKYAN, V. F., POGAREV, S. V.: Solid State Communication 55, 167 (1985)

SEIFERT, W., KOKOVIN, G. A.: unpublished results

STRINGFELLOW, G. B., HALL JR., H. T., BURMEISTER, R. A., J. Appl. Physics 46, 3006 (1975) a

STRINGFELLOW, G. B., WEINER, M. E., BURMEISTER, R. A.: J. Electron. Mater. 4, 363 (1975) b

STRINGFELLOW, G. B., HOM, G.: J. Electrochem. Soc. 124, 1806 (1977)

SUN, S. W., WESSELS, B. W.: J. Appl. Physics 57, 4616 (1985)

THEETEN, J. B., HOLLAN, L., CADORET, R.: Crystal Growth and Materials, ed. E. KALDIS and H. J. SCHEEL, Amsterdam 1977

WOLFORD, D. J., BRADLEY, J. A., FRY, K., THOMPSON, J.: Proc. 17th Internat. Conference on Semicond. Phys. San Francisco 1984

(Received January 29, 1987)

Authors' addresses:

Dr. S. SCHWETLICK, Doz. Dr. W. SEIFERT, Prof. Dr. E. BUTTER
Sektion Chemie der Karl-Marx-Universität Leipzig
Talstraße 35
Leipzig
DDR-7010

Dr. W. HÖRIG, Dr. R. PICKENHAIN, Dr. R. SCHWABE
Sektion Physik der Karl-Marx-Universität Leipzig
Linne-Straße 5
Leipzig
DDR-7010

Cryst. Res. Technol.	22	1987	8	1009–1014

D. SCHELLE, H.-J. TILLER

Sektion Chemie, Friedrich-Schiller-Universität Jena, DDR

Nb Redeposition on Si during Plasma Etching of Nb/SiO_2/Si Layers Investigated by RBS

The redeposition of electrode material (Ni, Cr, Al) and material of a Nb layer on silicon surface during plasma etching of Nb, SiO_2 and Si by a CF_4 plasma has been investigated by Rutherford Back Scattering spectrometry. It is shown that a steady state

$$(NbF_x)_{gas} \rightleftharpoons (Nb)_{ads} + xF$$

exists, what causes a Nb contamination of etched Si surfaces. Obviously this steady state concentration $(Nb)_{ads}$ is influenced by additional redeposition of non-etchable electrode material as Ni, Cr and Al.

Introduction

The contamination of CF_4-plasma etched Si surfaces by C_xF_y glowpolymers is a well known and often investigated phenomenon (MATSUMOTO, SUGANO) caused by the overall decomposition of CF_4

$$xCF_4 \rightleftharpoons C_xF_y + (4x - y)\,F\,.$$

Thereby the glowpolymers are formed during recombination and crosslinking of radicals and condensable reaction products. The glowpolymer formation is increased by F-consuming etching conditions (H containing gases) and decreased by oxidizing conditions (O-, F-, Cl addition).

The redeposition of material of the electrodes and the reactor wall depends on the bias voltage and the energy of sputtering ions therefore and superimposes the glowpolymer formation. This effect is used for the formation of metal containing glowpolymer (KAYE, DILKS; MARTINU, BIEDERMANN).

As a result of the metal embedment the sputtering propability is changed and the metal coverage is therefore related to the glowpolymer formation.

Glowpolymer formation and metal redeposition phenomena take place in principle from a theoretical point of view. The detections is last not least a analytical question. In the following the interaction between the redeposition of different metals (Nb, Cr, Ni) is investigated by Rutherford back scattering spectrometry.

The method is a very sensitive one for elements with $M > 30 \cdot (10^{12} - 10^{13})$ atoms/cm^2 (CHU et al. 1978).

Experimental

3 inches Si wafer partially covered with a 100 nm thick Nb layer are etched in a planar reactor with electrodes 12.8 cm in diameter, 3.5 cm separated and both isolated against the reactor walls. The used electrodes were water cooled Cr or Ni disks. The CF_4 plasma

etching parameters were the following:

CF_4 pressure — 13 Pa, power density — 0.40 W · cm^{-2};

CF_4 flow — 10 sccm, excitation frequency — 2.5 MHz.

For comparison experiments in an Ar plasma were carried out under similar conditions.

The RBS investigation have been done in an equipment described by GÖTZ et al. The energy E_0 of the helium ions was 1.4 MeV, the ion dosis — 10 μC, the scattering angle $\theta = 170°$, the backscattered ions were sampled by an barrier detector. The surface density of contamination $(Nt)_i$ has been computed in the surface approximation:

$$(Nt)_i = A_i/\Omega\sigma_i(E_0)\,Q$$

where Ω is the space angle spanned by the detector, Q the ion dosis and $\sigma_i(E_0)$ the cross-section of scattering He^+ ions with energy E_0. Figure 1 shows the detection limit of elements in dependence on the mass number for the RBS equipment we used.

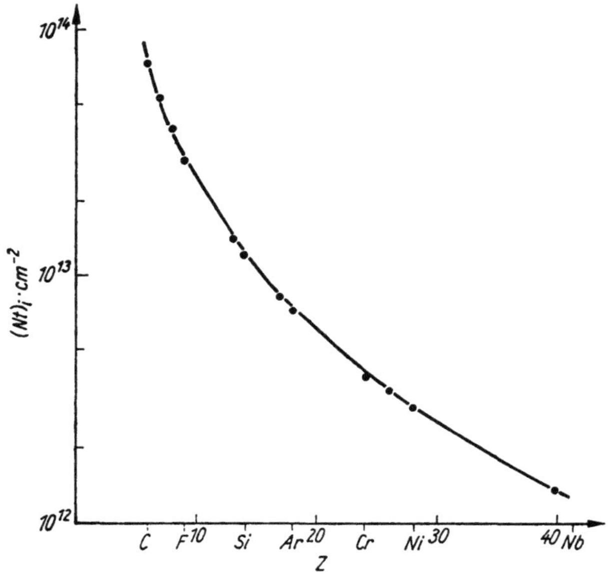

Fig. 1. Detection limit of elements by Rutherford backscattering spectrometry of the used equipment. 1.4 MeV He^+, 10 μl, $\theta = 170$ °, $\Omega = 2.58$ msr

Results and Discussion

Figure 2 shows an RBS spectrum of an silicon sample etched for 8 min in a CF_4 plasma. It is clearly seen the Ni scattering signal from the contaminated Si wafer, while Cr from the grounded counter electrode was not detected. Additional investigations of the sample (tilting angle $\theta = 55°$ and aligned spectra) show that there was an amorphous layer on top of the silicon containing nickel, fluorine, carbon and oxygen.

The spectrum on Figure 3 shows a more distinct evidence for the existence of carbon and fluorine using the channeling technique. That means the helium ion beam which impact on the surface is aligned with a major crystallographic direction of silicon. The carbon/fluorine ratio in the surface contaminent layer varies from 8 in

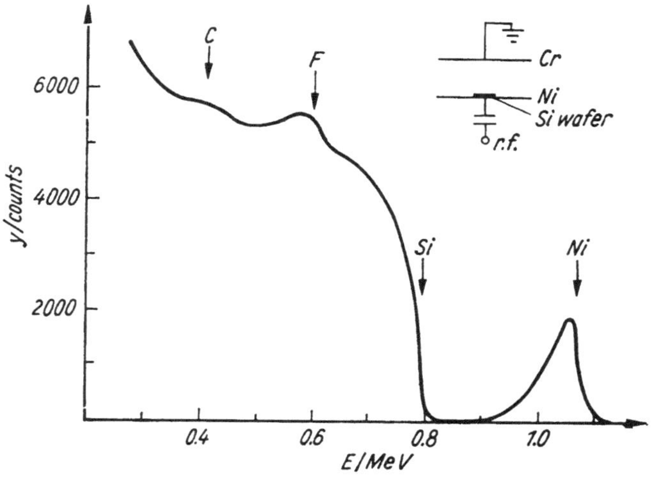

Fig. 2. RBS spectrum of a Si wafer etched within a planar reactor on a Ni coated r.f. electrode by a CF_4 plasma (etching conditions see: experimental)

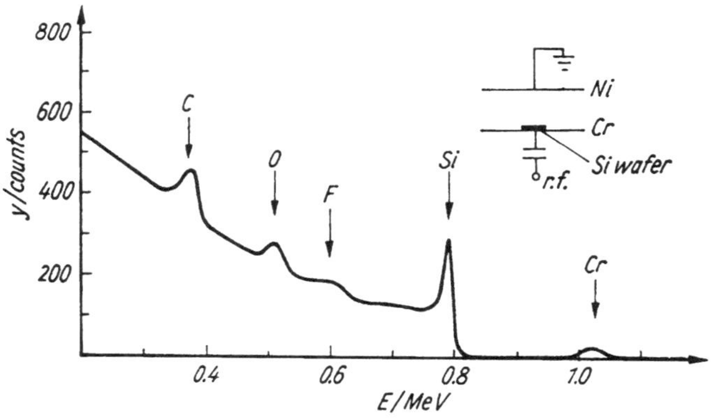

Fig. 3. Aligned RBS spectrum of a Si wafer etched within a planar reactor on a Cr coated r.f. electrode by a CF_4 plasma (etching conditions see: experimental)

this case to about 1 in the highly metal contaminated case (nickel, aluminium after long plasma exposition).

Comparing the redeposition of Ni and Cr on the Si wafer in Ar (sputter redeposition) and CF_4 plasma significant differences are observed. The amount of redeposited Ni is proportional to the time (Fig. 4), whereby the redeposition within the CF_4 plasma is nearly twice the amount obtained in the Ar plasma. The sputtering efficiency of Ar and CF_4 is similar. Therefore the embedment of Ni within the C_xF_y glowpolymer layer should be the reason for Ni accumulation on the Si surface. Ni is not etchable by F containing plasmas ($p_{NiF_2} = 0.13$ Pa, $T = 1033$ K (Efimov 1983). The Ni embedment is also proved by the asymmetrical profile of Ni signal seen on the RBS spectrum on Figure 2.

The Cr redeposition by an Ar plasma is quantitativly similar to that of Ni (Fig. 5). But the amount remains constant with time. This may be an impurity effect of C_xF_y

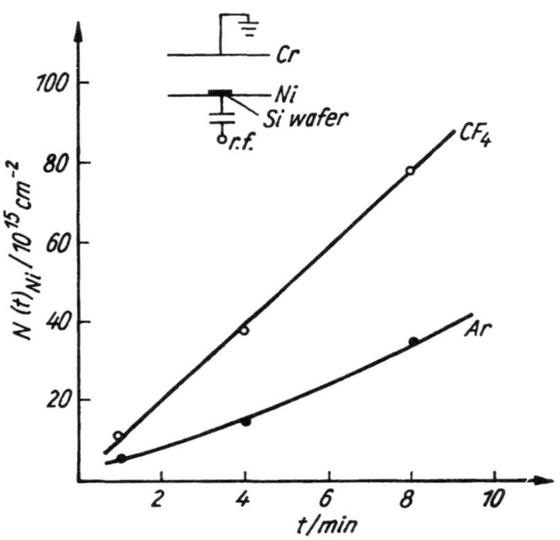

Fig. 4. Ni contamination of a Si wafer on a Ni coated r.f. electrode etched by a CF$_4$ plasma and sputter etched by an Ar plasma in dependence on etching time (etching conditions see: experimental)

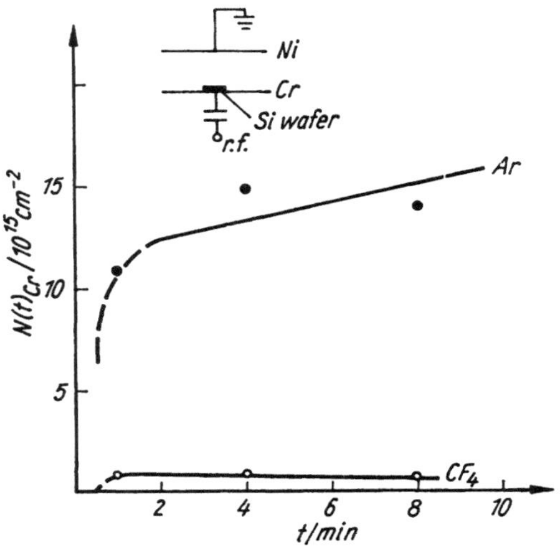

Fig. 5. Cr contamination of a Si wafer on a Cr coated r.f. electrode etched by a CF$_4$ plasma and sputter etched by an Ar plasma in dependence on etching time (etching conditions see: experimental)

residual gas within the reactor equipment. The pure CF$_4$ plasma produces a small Cr redeposition only which is constant with time and (Fig. 5) more than one order of magnitude below that of Ni. Whereas the Cr fluorides have a low vapour pressure ($p_{CrF_2} = 0.13$ Pa, $T = 1121$ K; $p_{CrF_3} = 0.13$ Pa, $T = 1087$ K) (EFIMOV 1983). the CrOF$_x$ formed by oxygen residual gas is a more volatile compound. That means Cr is of limited etchability by CF$_4$ (CCl$_2$F$_2$/O$_2$ mixture is an usual etching gas for Cr) depending on the plasma gas condition. The experimental result of Cr redeposition in dependence on time gives some hints that the Cr embedment into the C$_x$F$_y$ glow-polymer is low and only an equilibrium coverage of Cr on the substrate surface is formed.

Nb is etchable by CF$_4$ with etch rates of 80—120 nm/min in the used planar reactor and by the described conditions. The volatile etch product is NbF$_5$ with a vapour pressure $p_{NbF_5} = 1.33$ Pa, $T = 298$ K. The Nb contamination was studied with the aid of a small Nb area (1 cm^2) (600 nm Nb on Si) positioned near the wafer (Fig. 6). The Si etch rate was 100 nm/min. After 6 min plasma etching Nb was detected on the whole wafer. The amount was below that of Cr. The Nb contamination distribution across the wafer is seen from Figure 6.

It should be expected a decrease of Nb contamination from point A to B. Instead of this an increase of the Nb contamination to the opposite wafer edge is detected. The Nb-contamination correlates to the Cr contamination caused by the electrode material.

The etching of Nb (similar results may be expected for other materials, too) follows an equilibrium reaction:

$$\mathrm{Nb} + 5\,\mathrm{F} \rightleftharpoons \mathrm{NbF_5} \xrightarrow{\text{transport}} (\mathrm{Nb})_{ads} + 5\,\mathrm{F}$$
$$\mathrm{NbF_5} \xrightarrow{\text{transport}} (\mathrm{NbF}_x)_{ads} + (5 - x)\,\mathrm{F}\,. \qquad (1)$$

That means a deposition of Nb or nonvolatile NbF$_x$ on Nb-free areas of the reactor and the wafer.

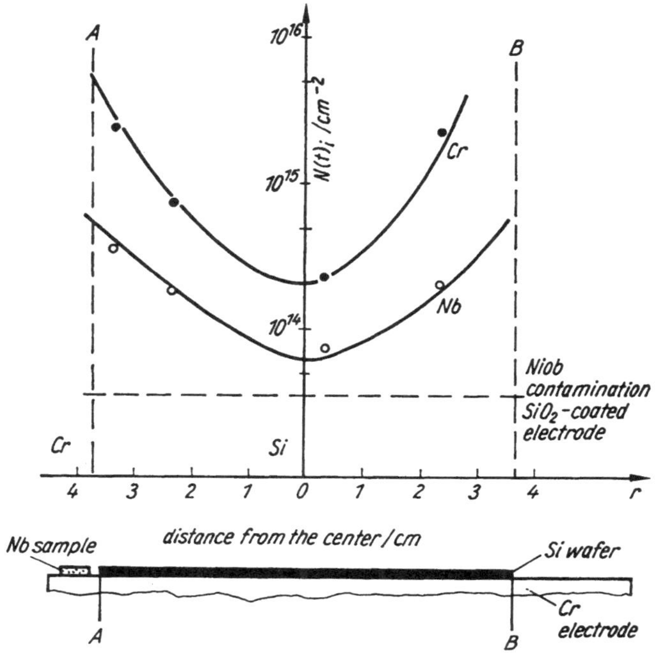

Fig. 6. Cr and Nb contamination ditsribution on a 3 inches Siwafer on a Cr r.f. electrode etched by a CF_4 plasma (etching conditions see: experimental)

The Cr distribution on the wafer should be symmetrically because the wafer is surrounded by the Cr electrode. Therefore the conclusion of an interaction or influence of the Cr redeposition on the Nb re-adsorption is possible. The Cr can cover the Nb or can form mixed products as fluorides, oxyfluorides or oxides.

The role of oxide formation is proved by the Nb contamination on Si compared to SiO_2. After an CF_4 etching time of 4 min an average Nb contamination on the edge of a Si wafer (experimental arrangement see Figure 6) amounted to $2.9 \cdot 10^{14}$ cm^{-2} and on the oxidized wafer $4.8 \cdot 10^{14}$ cm^{-2}. An reversible reactive desorption of Nb corresponding to equation (1) is inhibited by a $(NbO_x)_{ads}$ formation.

Table 1

Average contamination of a 3 inches Si wafer after CF_4 etching on an r.f. electrode covered with different materials. Etching conditions as described earlier

r.f. electrode material	etching time/min	C/cm^{-2}	F/cm^{-2}	Me/cm^{-2}	C/F ratio
SiO_2	8	$1.7 \cdot 10^{16}$	$5.4 \cdot 10^{15}$	—	3.1
Cr	8	$1.1 \cdot 10^{16}$	$3.4 \cdot 10^{15}$	$1.9 \cdot 10^{14}$	3.2
Ni	1	$4.8 \cdot 10^{16}$	$1.3 \cdot 10^{6}$	$1.2 \cdot 10^{16}$	3.6
	4	$5.4 \cdot 10^{16}$	$4.4 \cdot 10^{6}$	$4.1 \cdot 10^{16}$	1.2
	32	$7.3 \cdot 10^{16}$	$2.1 \cdot 10^{17}$	$1.7 \cdot 10^{17}$	0.3

A quantitative estimate of the C/F ratio in dependence on the metal contamination proves the assumption of an MeF_x imbedment within the C_xF_y glowpolymer formation to be correct (Tab. 1). Ni contaminated samples show an increase of F contamination. Assuming an C/F ratio of about 3 for the pure glowpolymer a NiF_x formation with $x \lessapprox 1$ results from the estimation. The glowpolymer layer is formed during the first etching phase arriving a value with a small increase with time only.

The degree of glowpolymer formation also is related to the metal contamination. Strongly metal contaminated surfaces (Ni) show a enhanced glowpolymer formation as seen from Table 1 if the C contamination for Cr and Ni electrodes is compared.

Conclusions

The contamination of CF_4 etched Si wafers with electrode materials (Cr, Ni), layer material (Nb) and glowpolymer (C_xF_y) shows a very complex interaction between the contaminated materials. Etchable materials as Nb are contaminated via a decomposition equilibrium of the volatile NbF_5. The contamination is enhanced by an oxide formation and is influenced by the contamination of other metals as Cr for instance.

The contaminated Ni partially is deposited in form of NiF_x what enhances the glowpolymer formation. This can be understood by an inhibition of the reactive desorption of C_xF_y glowpolymer within the steady state

$$\begin{array}{l} nCF_4 \rightleftharpoons C_nF_s + (4n - s)\,F \\ \qquad\quad \upharpoonleft\downharpoonright \\ \qquad\quad (C_xF_y)_{polymer} + iF + C_{(n-x)}F_{(s-(y+i))}\,. \end{array}$$

References

CHU, W. U., MAYER, J. W., NICOLET, N. A.: Backscattering Spectrometry, New York 1978

EFIMOV, A. I.: Neorganiceskich Sojedinenij Spravocnik, Leningrad 1983

GÖTZ, G., SCHWABE, F., TREFF, H. J.: Radioanalyt. Chem. **28**, 43 (1975)

KAYE, E., DILKS, A.: Thin Solid Films **78**, 309 (1981)

MARTINU, L., BIEDERMANN, H.: ISPC-7, Eindhoven 1985, p. 1315

MATSUMOTO, H., SUGANO, T.: J. Electrochem. Soc. **129**, 2823 (1983)

(Received January 29, 1987)

Authors' address:
D. SCHELLE, Prof. Dr. H.-J. TILLER
Sektion Chemie der Friedrich-Schiller-Universität Jena
Lessingstr. 10
Jena
DDR-6900

Cryst. Res. Technol.	22	1987	8	1015–1022

A. Packter, Ms. K. S. Panesar

Chemistry Department, North-East London Polytechnic, London

The Precipitation of Aluminium Hydroxocarbonates Powders from Aqueous Solution: Precipitate Compositions and Precipitation Mechanisms

A series of aluminium hydroxocarbonate hydrates were prepared by precipitation from aluminium nitrate solution, with five sodium hydrogen carbonate-sodium carbonate solutions of different *p*H, at ambient temperature. The course of precipitation was monitored by *p*H measurement and the final precipitate compositions were determined by chemical analysis, infra-red spectrophotometry, X-ray diffraction and thermal analysis.

Precipitation generally occurred through three stages, primary precipitation of materials with low carbonate content at low *p*H with evolution of carbon dioxide, their dissolution to form hydroxcarbonato anions and then secondary reprecipitation of the final products at higher *p*H. These materials were mixtures of polyhydroxoaluminium carbonate hydrates of general composition $Al_n(OH)_{3n-2}CO_3 \cdot hH_2O$ (where $n = 2, 4, 6$ and $h = 6-8$); their CO_3 content increased with increasing *p*H and carbonate anion content of the precipitant solution.

1. Introduction

The aluminium hydroxocarbonates are of some general theoretical interest to inorganic chemists (Mellor) and also have some useful applications as anti-acid pharmaceuticals (Anon; Beckman; Haynes; Martin) and as precursors for the preparation of alumina adsorbents, catalysts and refractories (Clarke; Mansour, Ryabin). The preparation and structures of the divalent alkaline-earth and transition metal hydroxocarbonates have been studied in some detail but there have been few investigations of the corresponding aluminium salts; Mellor has reviewed the earlier preliminary studies while more recently Schvartsman has investigated the preparation of materials of composition $Al_xOH_{3x-2}CO_3 \cdot hH_2O$. Duffin has reported a study of the rare mineral Scarborite of composition $Al_{14}(OH)_{36}(CO_3)_3 \cdot 16\,H_2O$. A more detailed and systematic study would then be valuable.

In this work, a series of aluminium hydroxocarbonates were prepared by precipitation from aluminium nitrate solution with five different sodium hydrogen carbonate-sodium carbonate solutions at ambient temperature. The course of precipitation was monitored by pH titration and the final precipitate compositions were analysed by chemical analysis, infra-red spectrophotometry, X-ray diffraction and thermal analysis. The complex precipitation mechanisms are discussed.

2. Experimental

2.1. Materials and solutions

Aluminium nitrate, sodium hydrogen carbonate, sodium carbonate and sodium hydroxide were British Drug House AnalaR reagents. Stock solutions were prepared in CO_2-free distilled water; aluminium concentrations were checked by complexometric analysis and carbonate and hydroxide concentrations were checked by titration against standard hydrochloric acid.

2.2. Preparation

A series of aluminium hydroxocarbonates were prepared by precipitation from aluminium nitrate solution (0.2 to 0.5 M) with different carbonate solutions (1M and 2M);

precipitate PA — with $NaHCO_3$ solution (*p*H 8.1),
precipitate PB — with 4/1 $NaHCO_3$–Na_2CO_3 solution (*p*H 9.4),
precipitate PC — with 1/1 $NaHCO_3$–Na_2CO_3 solution (*p*H 10.3),
precipitate PD — with Na_2CO_3 solution (*p*H 11.5) and
precipitate PE — with 10/1 Na_2CO_3–NaOH solution (*p*H 12.3).

Generally the cation solution was stirred magnetically in wide beakers (at 200 r.p.m.) and the precipitant solution was added from a burette; the solution temperature was 20 °C.

In preliminary runs (100 cm³ cation solution), the precipitant was added in increments to well past the equivalent volume for complete metathesis; the solution *p*H after each increment was monitored with a glass electrode and accurate *p*H Meter standardised for the range 4 to 13 with a series of buffer solutions. These *p*Hs were corrected for the effects of volume change during titration.

These precipitations were then repeated (500 cm³ cation solution) and the precipitant was now added to slight excess only. The precipitate slurries were washed free of electrolytes by decantation and several reslurryings, filtered, collected and dried over silica gel to constant weight at ambient temperature (120 h), at 60 °C (72 h) and at 100 °C (48 h). The dried powders were stored in sealed bottles over silica gel in desiccators.

2.3. Precipitate compositions

2.3.1. Chemical analysis

Precipitate powder (0.3 g) was dissolved in dilute hydrochloric acid and the solution was made up to standard volume. Aluminium was determined in aliquot volumes by titration against standard sodium EDTAate solution.

Carbonate content was determined with a Schrotter CO_2 Alkalimeter (Griffin and George model GAV 700 T). — Total weight loss on ignition was determined on samples (0.3 g) heated to constant weight at 700 °C. — Empirical compositions were then estimated.

2.3.2. Infra-red spectrophotometry

Precipitate powder (3 mg) was preground in an agate mortar to c 2 µm size and dispersed in anhydrous KBr (300 mg) and discs were prepared under 1.5 ton pressure. The infra-red spectra were recorded on a Perkin Elmer I.R. Grating Spectrophotometer against pure KBr discs at 4000 to 400 cm^{-1}.

2.3.3. X-ray diffraction

Powder samples were preground to c 5 µm size. Powder photographs were obtained with a Philips Powder Camera designed for the Straumanis method. The *d* values for the main strong and medium bands were calculated from the spacings and were compared with Duffin's values for the aluminium hydroxocarbonate mineral Scarbonite and the ASTM values for different aluminium hydroxides.

2.3.4. Thermal analysis

Precipitate powder (5 to 8 mg) was heated in nitrogen from ambient temperature to 773 K in a Differential Scanning Calorimeter at constant rate 8 °C min^{-1}. The dH/dT vs. T thermograms were recorded and the initial and peak temperatures for the main thermal events were noted.

3. Results

3.1. Course of Precipitation

A series of aluminium hydroxocarbonates was prepared from aluminium nitrate solutions (0.3 to 1 M) by precipitation with different sodium hydrogen carbonate-carbonate and sodium carbonate-hydroxide solutions at ambient temperature. The course of these precipitations was first studied by potentiometric titration and typical pH vs V_{CO_3} plots are presented in Figure 1. These showed little similarity to the

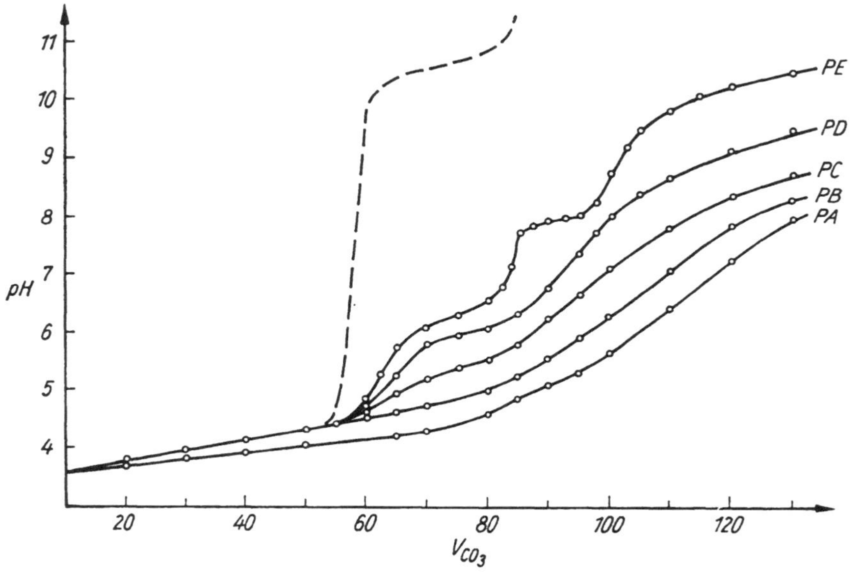

Fig. 1. Precipitations of aluminium hydroxocarbonates from 0.5 M aluminium nitrate solution with different carbonate solutions: pH titration plots. A) precipitant HCO_3^-, B) precipitant 4/1 HCO_3^-–CO_3^{2-}, C) precipitant 1/1 HCO_3^-–CO_3^{2-}, D) precipitant CO_3^-, E) precipitant 10/1 CO_3^{2-}–OH^-

corresponding plot for aluminium hydroxide precipitation (dashed lines) described in our previous work (PACKTER, PANESAR). Precipitation first occurred with vigorous evolution of carbon dioxide at pHs from 4 to 5; then with further addition of carbonate solution, the primary products started to redissolve while the final precipitates were formed concomitantly at pHs from 6 to 9. The pH vs V_{CO_3} plots indicated two to three precipitation events with inflexions at pH_1, pH_2 and pH_3 at precipitant volumes V_1, V_2, and V_3. The pH_1 values varied from 4.4 to 5.5, the pH_2 values from 5 to 7 while the pH_3 values varied from 7.5 to 8.8. Generally also

$$V_1 = 0.9 \text{ to } 1.0 V_e ,$$

$$\Delta V_2(V_2 - V_1) = 0.2 - 0.3 V_e ,$$

$$\Delta V_{13}(V_3 - V_2) = 0.2 V_e$$

where V_e is the equivalent volume of carbonate solution required for complete precipitation of the aluminium cation.

3.2. Precipitate compositions

3.2.1. Chemical analysis

The compositions of the precipitates prepared by different methods were then determined by various physical and chemical techniques. The compositions of the materials dried at ambient temperature and at 100 °C are collated below.

The CO_3/OH composition ratios tend to increase with increasing CO_3^{2-} content of the precipitant solution.

Table 1

Aluminium hydroxocarbonate compositions

	precipitant	precipitate compositions	
		dried at 20 °C	dried at 100 °C
PA	HCO_3^-	$Al_{3.9}OH_{9.7}CO_3 \cdot 8\ H_2O$	$Al_{3.7}OH_{9.7}CO_3$
PB	4/1 HCO_3^-–CO_3^{2-}	$Al_{3.3}OH_{7.7}CO_3 \cdot 7\ H_2O$	$Al_{3.3}OH_{7.7}CO_3$
PC	1/1 HCO_3^-–CO_3^{2-}	$Al_{2.9}OH_{6.7}CO_3 \cdot 7\ H_2O$	$Al_{2.7}OH_{6.7}CO_3$
PD	CO_3^{2-}	$Al_{2.9}OH_{6.7}CO_3 \cdot 7\ H_2O$	$Al_{2.7}OH_{6.7}CO_3$
PE	10/1 CO_3^{2-}–OH	$Al_{2.5}OH_{5.5}CO_3 \cdot 6\ H_2O$	$Al_{2.5}OH_{5.5}CO_3$

3.2.2. Infra-red spectrophotometry

The infra-red spectra of the different precipitates are presented in Figure 2; the spectrum of the mineral scarborite (Duffin) is included for comparison. These spectra confirmed the low carbonate/hydroxide composition ratios.

3.2.3. X-ray diffraction

The detailed results are available in Ms Panesar's Research Report (Pansear). Precipitate PA showed only the strong 4.91 Å line a weak 4.71 Å line and the strong 4.16 Å line from scarborite (Duffin). Precipitates PB, PC and PD showed no 4.91 Å line, the strong 4.72 Å line, several lines from the 4.16 — 3.73 Å group and several lines from the 2.86 — 2.63 Å group of scarborite. Precipitate PE showed a strong 5.98 Å line and several lines from the 4.16 — 3.73 Å group and the 2.86 — 2.63 Å group of scarborite. These results indicate that the precipitates were mixtures of two (or more) aluminium hydroxocarbonates and **not** mixtures of aluminium hydroxide with some high carbonate material $AlOHCO_3$.

3.2.4. Thermal analysis

Typical thermograms (for decompositions at 8 °C min^{-1}) are presented in Figure 3; the thermogram for gibbsite decomposition is included for comparison. None of the thermograms showed the $Al(OH)_3$ decomposition (to AlOOH) with high ΔH peak at $T_p = 590$ K (Brindley; Packter, Dhillon) but they all showed series of overlapping peaks with peak temperatures $T_{p1} = 670-680$ K, $T_{p2} = 735-750$ K and $T_{p3} = 750-765$ K. These confirm the results of the X-ray diffraction data; the thermograms indicate a continuous series of decompositions of hydroxocarbonates to oxohydroxocarbonates, then to oxocarbonates (and finally to aluminium oxides) similar to the series of decompositions noted for magnesium hydroxocarbonate (Visinti).

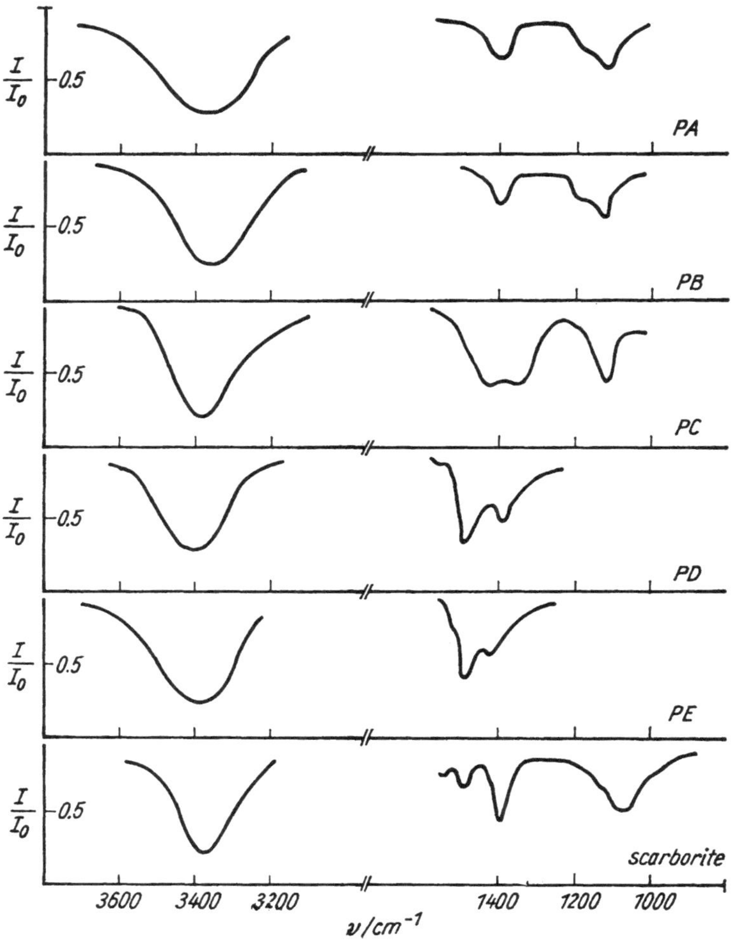

Fig. 2. Infra-red spectra of aluminium hydroxocarbonates

3.2.5. Precipitate compositions

Generally then the experimental studies suggest that the precipitates PA to PE are mixtures of polyhydroxoaluminium carbonate hydrates of the following compositions:

$$[Al_3(OH)_8]\,CO_3\,[(OH)_8Al_3] \cdot hH_2O \text{ — 'AHC 6'},$$

$$[Al_2(OH)_8]\,CO_3\,[(OH)_8Al_2] \cdot jH_2O \text{ — 'AHC 4'}$$

and

$$[Al(OH)_2]\,CO_2\,[(OH)_2Al] \cdot kH_2O \text{ — 'AHC 2'}$$

where $h \cong 10$, $j \cong 8$, and $k \cong 6$.

Precipitate PA is probably a mixture of about 48 percent 'AHC 6' and 52 percent 'AHC 2' or alternatively a mixture of about 95 percent 'AHC 4' and 5 percent 'AHC 2'.

Precipitate PB is probably a mixture of about 65 percent 'AHC 4' and 35 percent 'AHC 2'.

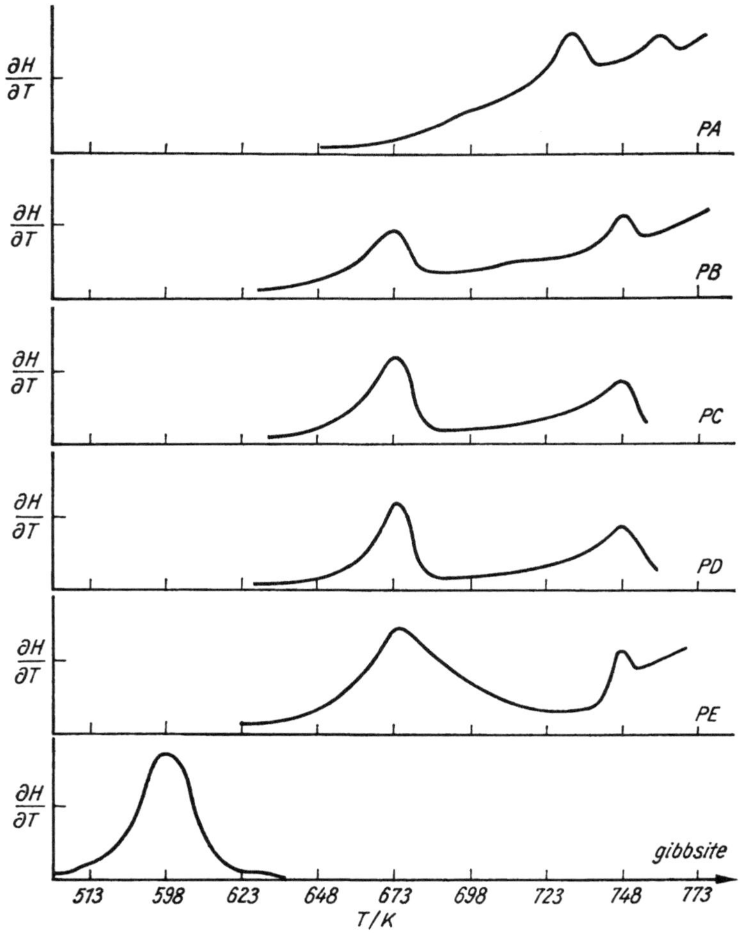

Fig. 3. Differential scanning calorimetry thermograms for the decomposition of aluminium hydrocarbonates in nitrogen at 8 °C · min⁻¹

Precipitates PC and PD are probably mixtures of about 55 percent 'AHC 4' and 45 percent 'AHC 2'.

Precipitate PE is probably a mixture of about 75 percent 'AHC 2' with 25 percent 'AHC 4'.

4. Discussion

Primary precipitations at low *p*H: The preparation of aluminium hydroxocarbonates from aluminium solutions may occur through a series of primary precipitation, dissolution and secondary reprecipitation processes. Aquoaluminium cations are generally hydrolysed in low *p*H aqueous solutions to a series of hydroxocations and polyhydroxocations (BAES; GIMBLETT). On addition of different hydrogen carbonate-carbonate solutions to aluminium nitrate systems, the hydrogen carbonate anions would react with the acidic hexaaquoaluminium cations to form more basic hydroxoaluminium cations — with evolution of carbon dioxide — and then the primary aluminium hydroxocarbonate precipitates. The main equilibria might then be as follows:

$$[Al(H_2O)_6]^{3+} + HCO_3^- \rightleftharpoons [AlOH(H_2O)_5]^{2+} + H_2O + CO_2$$

then

$$6[AlOH(H_2O)_5]^{2+} + 12\,HCO_3^- \rightleftharpoons Al_6(OH)_{16}(H_2O)_hCO_3 + 11\,CO_2 + (31 - h)\,H_2O$$

and

$$4[AlOH(H_2O)_5]^{2+} + 8\,HCO_3^- \rightleftharpoons Al_4(OH)_{10}(H_2O)_jCO_3 + 7\,CO_2 + (21 - j)\,H_2O;$$

also

$$[AlOH(H_2O)_5]^{2+} + HCO_3^- \rightleftharpoons [Al(OH)_2(H_2O)_4]^+ + H_2O + CO_2\,,$$

then

$$6[Al(OH)_2(H_2O)_4]^+ + 6\,HCO_3^- \rightleftharpoons Al_6(OH)_{16}(H_2O)_hCO_3 + 5\,CO_2 + (25 - h)\,H_2O$$

and

$$4[Al(OH)_2(H_2O)_4]^+ + 4\,HCO_3^- \rightleftharpoons Al_4(OH)_{10}(H_2O)_jCO_3 + 3\,CO_2 + (17 - j)\,H_2O\,.$$

Dissolution: On addition of further hydrogen carbonate-carbonate solution, the primary hydroxocarbonate precipitates may partially redissolve (in excess carbonate) to form polyhydroxocarbonatoaluminate anions (GROSCHENS; TOMILOV). The equilibria for the formation of these complex anions might be as follows:

$$Al_6(OH)_{16}(H_2O)_hCO_3 + CO_3^{2-} \rightleftharpoons [Al_6(OH)_{16}(H_2O)_h(CO_3)_2]^{2-}$$

and

$$Al_4(OH)_{10}(H_2O)_jCO_3 + CO_3^- \rightleftharpoons [Al_4(OH)_{10}(H_2O)_j(CO_3)_2]^{2-}$$

Secondary precipitations at intermediate *p*H: The hydroxocarbonatoaluminate anions must then disproportionate to form the secondary aluminium hydroxocarbonate precipitates. The equilibria would probably be as follows:

$$[Al_6(OH)_{16}(H_2O)_h(CO_3)_2]^{2-} + (j + k - h)\,H_2O \rightleftharpoons$$
$$\rightleftharpoons Al_4(OH)_{10}(H_2O)_jCO_3 + Al_2(OH)_4(H_2O)_kCO_3 + 2\,OH^-$$

and

$$[Al_4(OH)_{10}(H_2O)_j(CO_3)_2]^{2-} + (2k - j)\,H_2O \rightleftharpoons 2Al_2(OH)_4(H_2O)_kCO_3 + 2\,OH^- + 2\,OH^-\,.$$

Precipitations with hydrogen carbonate-carbonate buffer solutions then give mainly the primary precipitates with 5 to 45 percent secondary precipitates: precipitations with carbonate-hydroxide solutions on the other hand give predominately secondary precipitates.

References

Anon: Fr. Patent 483 (1962)
BAES, C. F. Jr.: The Hydrolysis of Cations, New York 1976, Ch 6, 112
BECKMAN, S. M.: U.S. Patent 2945783 (1960)
BRINDLEY, G. W.: Amer. Mineral. **46**, 771 (1961)
CLARKE, R. G.: U.S. Patent 3141729 (1964)
DUFFIN, W. J.: Mineral. Mag. **32**, 353 (1960)
GIMBLETT, M. G. R.: Inorganic Polymer Chemistry, London 1963, Ch 3.2., 80
GROSCHENS, D.: C.R. Acad. Fr. **209**, 313 (1936)
HAYNES, H. G.: Pharm. J. **157**, 154 (1946)
MANSOUR, N. A. L.: Interceram. **27**, 1 (1978).

MARTIN, J.: U.S. Patent 3272703 (1966)
MELLOR, J. W.: Comprehensive Treatise of Inorganic Chemistry, London 1954, Vol. 5, 358
PACKTER, A., DHILLON, H. S.: Kolloid Z. **239**, 508 (1970)
PACKTER, A., Ms. PANESAR, K. S.: Cryst. Res. Technol. **20**, 749 (1985)
PANESAR, K. S.: North-East London Polytechnic Research Report, 1987
RYABIN, S.: U.S.S.R. Patent 622759 (1978)
SCHVARTSMAN, B. Kh.: Zh. Prikl. Khim. **35**, 906 (1962)
TOMILOV, N. P.: Chem. Abstr. **68**, 45711i (1968)
VISINTI, B.: Gazz. chim. ital. **90**, 863 (1960)

(Received January 7, 1987)

Author's address:

c/o Dr. A. PACKTER
106 Howberry Road
Canons Park
Middlesex, England

Cryst. Res. Technol.	22	1987	8	1023–1030

D. KLIMM, P. PAUFLER

Karl-Marx-Universität Leipzig, Sektion Chemie,
Wissenschaftsbereich Kristallographie

Point Defects in GaP Single Crystals Investigated by Mechanical Damping

The dependence of mechanical damping of LEC (liquid encapsulated Czochralski) crystals of GaP:S was measured in the temperature range 300 ... 600 K and at frequencies near 70 and 140 kHz, respectively. In samples of different stoichiometry two peaks could be observed: Peak A at 350 K and peak B at 450 K. Peak A is attributed to the relaxation of point defects near the surface, peak B is probably due to Ga vacancies.

Die Abhängigkeit der mechanischen Dämpfung von schwefeldotierten LEC GaP-Kristallen wurde im Temperaturbereich 300 ... 600 K und bei Frequenzen von 70 bzw. 140 kHz gemessen. In Proben unterschiedlicher Stöchiometrie konnten zwei Maxima beobachtet werden: A bei 350 K und B bei 450 K. A wird der Relaxation von Punktfehlern in Oberflächennähe zugeordnet, B geht wahrscheinlich auf Ga-Leerstellen zurück.

1. Introduction

Sphalerite-type semiconducting III—V compounds play an important role in modern optoelectronics. It is well known that defects in the semiconducting material lower the efficiency and lead to degradation of devices. A variety of methods is used to characterize defects in semiconductor crystals, in most cases electrical and optical ones. Mechanical damping measurements, however, may well be an alternative since they are primarily sensitive to the strain field of defects.

From the point of view of crystal structure the following kinds of simple point defects can be expected in GaP: Vacancies of Ga and P, interstitials of both elements and antisite defects. Different authors propose the existence of nearly all of them (JORDAN et al. 1974a; TANAKA, SUKEGAWA, AKSENOV et al.; DZHAFAROV, KHUDYAKOV; KAUFMANN et al.; VERNER et al. 1979). A similar variety of intrinsic point defects should occur in other sphalerite-type semiconductors such as GaAs (CHAKRAVERTY, DREYFUS; LIN-CHUNG, REINECKE; TSUKADA et al.; SAMUELSON; FUJISAKI et al.). Moreover, extrinsic defects due to impurities and/or dopants should be taken into account.

Lowering dislocation density in such semiconductors mostly increases the concentration of "microdefects" (e.g. perfect and faulted dislocation loops and precipitations with diameters of about 100 nm; WAGNER et al.). Condensation of point defects is supposed to lead to formation of microdefects (DE KOCK).

Several authors already dealt with the internal friction in GaP (BOYLE, SLADEK; KIRILLOV et al.; KLIMM, PAUFLER; MITROCHIN et al.; PESIN et al.; VERNER et al. (1977, 1979). VERNER et al. reported on 3 peaks of damping at 403, 593, and 683 K for undoped n-GaP. They attributed them to divacancies. KLIMM and PAUFLER found two peaks in S-doped LEC GaP which were attributed to point defect relaxation.

The aim of the present paper is to supply further experimental data which can help to identify lattice defects which are responsible for those peaks.

2. Experimental

2.1 The piezoelectric oscillator

For the measurements we used piezoelectric oscillators as described earlier by MARX (Fig. 1).

The two quartz crystals are 18.5° X-cut (BERGMANN). They are cemented together with a two component epoxy resin. This resin also yields the link to the silica glass buffer rod. The link between buffer rod and sample is made of a mixture of talcum and waterglass or by a commercial high temperature cement.

Two different oscillators of that type were used to perform the measurements: One with quartz crystals of 20 mm length (resonance frequency about 140 kHz) and the other with quartz crystals of 40 mm length (resonance frequency about 70 kHz). During measurement the oscillator was suspended vertically by means of two copper wires (1 mm diameter) attached by silicon gum to the centres of the X-faces of quartz 2. The electrodes of quartz 1 and 2 as well as the electric connection between copper wires and electrodes were made of conducting silver paint (Fig. 2).

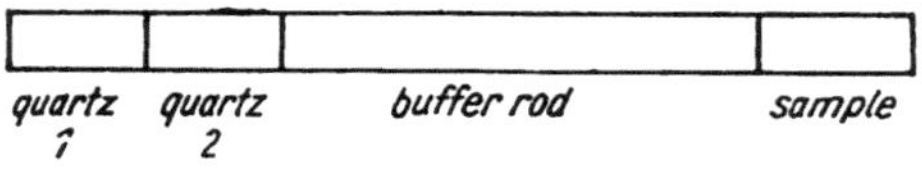

Fig. 1. Marx piezoelectric oscillator

Fig. 2. Suspension and contacts at the quartz crystals ▶

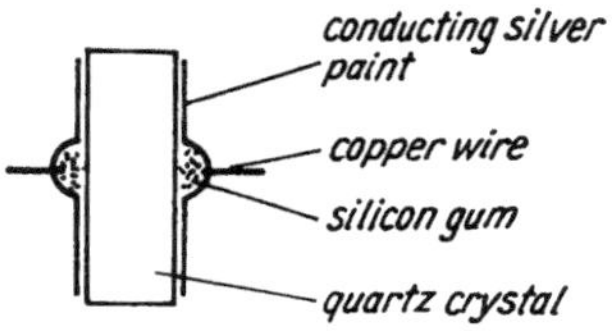

If a driving voltage V_d is applied to the X-faces of quartz 2 (frequency = resonance frequency of the oscillator), the whole piezoelectric oscillator is forced to longitudinal vibrations. These vibrations lead to a voltage V_g which can be measured at the electrodes of quartz 1. It can be shown that the ratio V_d/V_g is proportional to the damping of the whole oscillator (MARX). From a measurement of V_d and V_g one can easily obtain the logarithmic decrement of the sample alone as a function of temperature, if the sample is placed in a suitable furnace.

As the resonance frequency of the piezoelectric oscillator changes with temperature, the frequency of the applied voltage V_d has to be changed continuously during heating and cooling the sample. For this purpose we used the PLL (phase-locked loop) integrated circuit A 290 D (VEB Halbleiterwerk Frankfurt/Oder) to keep the phase angle between V_d and V_g at 90° during resonance. V_d was changed in the range of 0 ... 200 V which leads to amplitudes of vibration $\varepsilon = 0 \dots 6 \cdot 10^{-6}$. V_d and V_g were measured by analogue transistor voltmeters, the frequency of vibration by a frequency counter.

To reduce contamination of the sample during measurement, the whole piezoelectric oscillator was kept under a pressure of about 2 Pa. The sample was heated by an electric resistance furnace. A Pt–Pt/Rh thermocouple was placed near the sample to measure the temperature.

2.2. Samples

Samples were cut from a ⟨111⟩ LEC single crystal bar, provided by the VEB Gleichrichterwerk Stahnsdorf. The material was sulphur doped (carrier density $4 \cdot 10^{17}/cm^3$) and had an etch pit density of about $1.5 \cdot 10^5/cm^2$. Orientation and size of the samples (in mm) is shown in Figure 3.

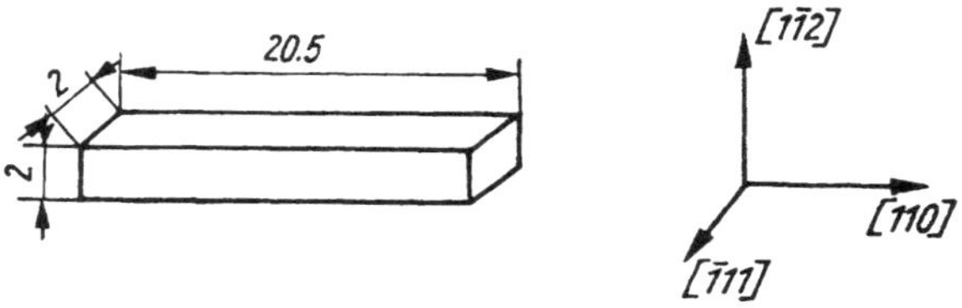

Fig. 3. Samples used for measuring internal friction

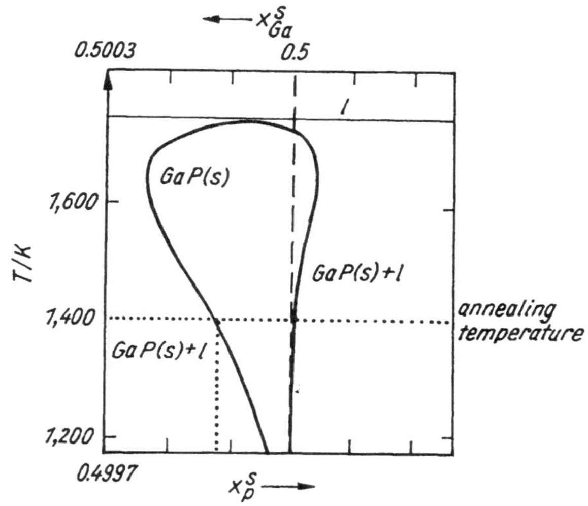

Fig. 4. Region of existence for GaP (JORDAN et al. 1974 a)

Before measurement of internal friction the chemical composition of some samples was changed by annealing them in sealed quartz glass ampoules under different pressures of P_2. According to the results of JORDAN et al. (1974b) at an annealing temperature of 1,400 K a phosphorus pressure of 250 kPa (duration: 3 months) or of 450 Pa (for 2 weeks only to keep evaporation losses small) is to be expected. That treatment should yield nearly stoichiometric GaP samples and samples with excess of Ga, respectively (Fig. 4). Thus the samples obtained should have different contents of the various kinds of intrinsic point defects (JORDAN et al. 1974b) (Tab. 1).

Table 1

Preparation and defect contents of samples with different stoichiometry

Composition	nearly stoichiometric	Ga-rich
designation of sample	GaP/P	GaP/Ga
annealing temperature	1 400 K	1 400 K
P_2 pressure	250 000 Pa	450 Pa
annealing time	3 months	2 weeks
density of Ga vacancies	$10^{18}/cm^3$	$10^{16}/cm^3$
density of P vacancies	$2 \cdot 10^{17}/cm^3$	$10^{19}/cm^3$
$(x_{Ga} - 0.5)$	≈ 0	10^{-4}

In contrast to JORDAN et al. (1974b) TANAKA and SUKEGAWA claim the existence of charged vacancies V_P^+ and V_{Ga}^-. Their concentration should be of the same order of magnitude as predicted for V_{Ga} and V_P by JORDAN et al. (1974b). Moreover, in the region of low P_2 pressure antisite defects Ga_P^{2-} should be present in approximately the same concentration as V_P^+.

3. Results

3.1. Damping measurements

In samples not annealed a maximum of internal friction was observed near 350 K (peak A). As already reported by KLIMM and PAUFLER it only occurred during the first heating cycle of the measurement. After that peak A could not be observed in the same sample, it reappeared, however, after damaging the surface by slightly abrading it with emery paper. If the abraded surface was etched before measurement of damping by aqua regia at 350 K for 5 minutes, peak A did not appear (Fig. 5).

As the shift between the temperatures of peak A at 70 kHz and at 140 kHz was too small to be interpreted in terms of activation energy, the peak width was used to estimate it (cf. NOWICK, BERRY). Thus an activation energy $E_A = (0.63 \pm 0.06)$ eV and a time constant $\tau_0 = 10^{-(15\pm1)}$ s entering the relaxation time $\tau = \tau_0 \exp(E_A/k_B T)$ were obtained. Time constants τ_0 of this order of magnitude are often observed for relaxation of point defects (DE BATIST). The observations indicate that point defects in the surface layer of the samples might be responsible for peak A.

In annealed nearly stoichiometric GaP/P samples another damping maximum has been observed at about 450 K (peak B). It reappeared in the same sample after one heating/cooling cycle several times, but the peak temperature varied slightly between first heating and the following measurements (Fig. 6). After about 5 measuring cycles

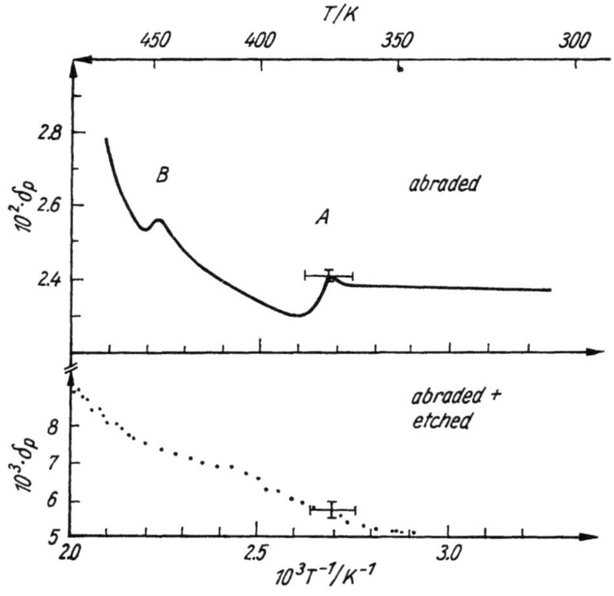

Fig. 5. Effect of abrading and etching the surface on the internal friction of the sample

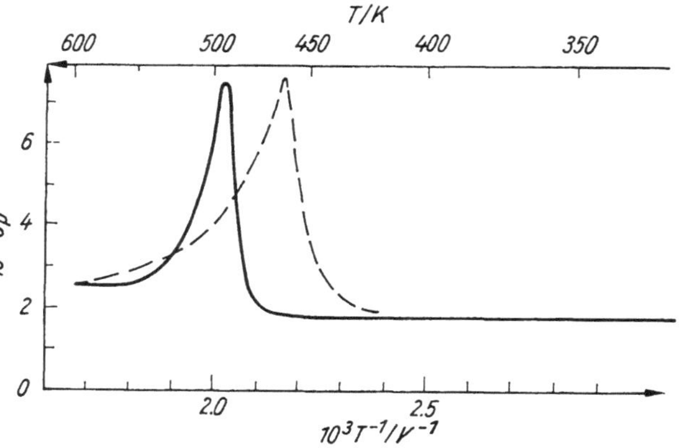

Fig. 6. Internal friction of a sample annealed under high P_2 pressure. ——— heating, – – – cooling

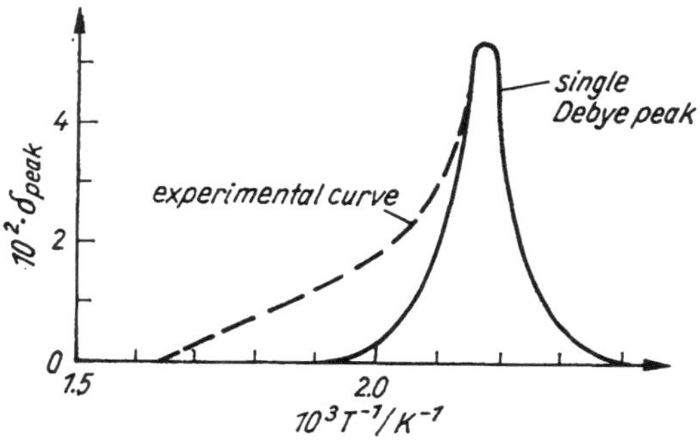

Fig. 7. Peak from Figure 6 without background

the peak disappears; recovery occurs after approximately 4 weeks. In contrast to that the Ga-rich samples did not show any damping maximum in the region from room temperature to 550 K.

If from the curve in Figure 6 background is substracted, it becomes evident that the damping peak B is not symmetric (Fig. 7). This could be due to a superposition of different relaxation processes. No effort has been made to work out the parameters of those less pronounced Debye peaks. From the maximum possible temperature-shift of peak B for different measuring frequencies E_A is estimated to be not less than 0.6 eV and τ_0 not to exceed 10^{-13} s.

Peak B is well pronounced in GaP/P samples. The fact, that the peak does not occur in (also annealed) GaP/Ga samples leads to the conclusion that it is not due to defects created by thermal stress during annealing. It seems that intrinsic point defects are responsible for the peak. They might be, e.g., Ga double-vacancies or larger clusters of Ga vacancies. Other experiments also suggest that intrinsic point defects could be the reason for the peak. Figure 8 shows the internal friction of a sample, irradiated with 1 MeV protons. That treatment produces mostly point defects in a surface layer about 25 μm thick. Both, peak A and B are visible.

An influence of plastic deformation upon internal friction of GaP could not be detected. There also does not occur a significant dependence of internal friction on vibration amplitude for $0 \leqq \varepsilon \leqq 5 \cdot 10^{-6}$. This indicates that dislocation should not

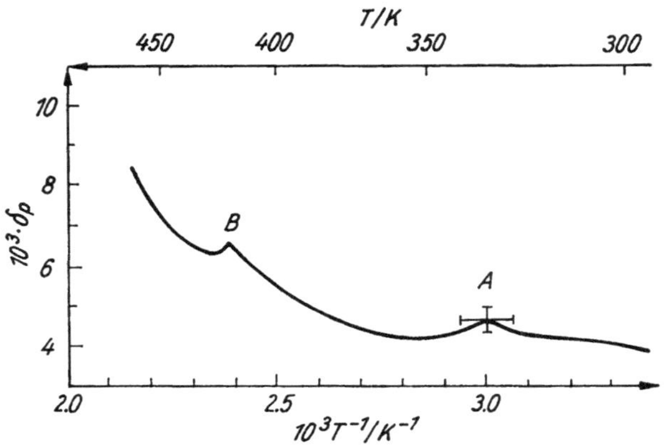

Fig. 8. Internal friction of a proton irradiated sample

play an important role for the damping of elastic waves under the conditions of the present investigation.

3.2. Other investigations

Electron transmission microscopy of the as-grown LEC GaP material shows long dislocation lines and about $2.6 \cdot 10^{11}$ microdefects/cm³ with diameters of 150 ... 200 nm. Annealing under high (250 kPa) phosphorus pressure changes the microdefects completely. The large (150... 200 nm) ones disappear and smaller ones with diameters in the region 40 ... 50 nm and densities of about 10^{12}/cm³ are observed. After measurement of internal friction the density of these microdefects does not change significantly, but small new microdefects with diameters of about 7 nm appear in approximately the same concentration (WAGNER). These very small microdefects seem to be newly created during the process of damping measurement. The process of creating new microdefects during the measurement of internal friction is accompanied by a decrease of mass density of the samples (Tab. 2, GaP/O as-grown material).

Table 2

Mass density before and after damping measurement (after KLÖSS)

Designation	ϱ/g cm^{-3}	rel. change dur. meas.	exp. error
GaP/O (before meas.)	4.1285	−0.07%	±0.024%
GaP/O (after meas.)	4.1257		
GaP/P (before meas.)	4.1260	−0.25%	
GaP/P (after meas.)	4.1157		

It can be assumed that the measurement of internal friction creates vacancies in the material, which agglomerate to those small microdefects of about 7 nm diameter. This could explain the lowering of mass density. Positron annihilation measurements show an increase of the S parameter after the measurement of annealed samples (EHRHARDT). This fact also can be interpreted in terms of increased vacancy concentration in the material.

No differences in infrared absorption between annealed and non-annealed samples (which could reveal a change of point defect household) have been observed as well as before and after the measurement of internal friction (SOBOTTA).

4. Conclusion

At measuring frequencies in the 100 kHz region two internal friction peaks could be observed in GaP. The low temperature peak A disappeared after the first measurement, whereas the high temperature peak B remained relatively stable. The occurrence of peak B strongly depends on deviation from stoichiometry of the sample. Similar results were obtained with GaAs by CHAKRAVERTY, DREYFUS.

VERNER et al. (1977) and VERNER et al. (1979) observed in GaP at measuring frequencies in the 100 kHz region 3 peaks, which they attributed to point defect relaxations. We conclude that peak A observed by them is of the same origin than our peak B, e.g., due to relaxations of Ga-divacancies or higher aggregated point defect clusters.

We are grateful for the performance of various measurements and investigations by C. Ascheron, G. Klöß, S. Müller, H. Sobotta, G. Wagner (Karl-Marx-Universität Leipzig) and by R. Krause (Martin-Luther-Universität Halle/Wittenberg). We also thank H. Fröbel, D. Oppermann, G. Wegener and W. Wetzel for technical help. Moreover, we would like to appreciate the valuable discussions with N. A. Tjapunina (Lomonossov University Moscow).

References

AKSENOV, V. F., ERMALOV, O. N., IGNATKINA, R. S.: Neorg. Materialy **20**, 1429 (1984)

BATIST, R. DE: Internal friction of structural defects in crystalline solids; Amsterdam, London, New York 1972

BERGMANN, L.: Der Ultraschall und seine Anwendung in Wissenschaft und Technik; Stuttgart 1954

BOYLE, W. F., SLADEK, R. J.: Internal friction and ultrasonic attenuation in crystalline solids I; Berlin, Heidelberg, New York 1973

CHAKRAVERTY, B. K., DREYFUS, R. W.: J. Appl. Physics **37**, 631 (1966)

DZHAFAROV, T. D., KHUDYAKOV, S. V.: phys. stat. sol. (a) **63**, 431 (1981)

EHRHARDT, N.: Diploma work; Martin-Luther-Universität Halle 1986

FUJISAKI, Y., TAKANO, Y., ISHIBA, T., SAKAGUCHI, H., ONO, Y.: Jap. J. Appl. Physics **24**, L899 (1985)

JORDAN, A. S., von NEIDA, A. R., CARUSO, R., KIM, C. K.: J. Electrochem. Soc. **121**, 153 (1974a)

JORDAN, A. S., CARUSO, R., von NEIDA, A. R., WEINER, M. E.: J. Appl. Physics **45**, 3472 (1974b)

KAUFMANN, U., SCHNEIDER, J., RÄUBER, A.: Appl. Phys. Lett. **29**, 312 (1976)

KIRILLOV, V. I., PRIBYLOV, N. N., REMBEZA, S. I., SPIRIN, A. I., TESLENKO, V. V.: Fiz. Tverd. Tela **24**, 1494 (1982)

KLIMM, D., PAUFLER, P.: in: Tarcie wewnetrzne i opóźnienie magnętczne w ciałach stałych; Katowice 1985, p. 203

KLÖSS, G.: Thesis, Karl-Marx-Universität Leipzig 1986

DE KOCK, A. J. R.: in: Handbook on semiconductors III; Amsterdam, New York, Oxford 1980

LIN-CHUNG, P. J., REINECKE, T. L.: Phys. Rev. B **27**, 1101 (1983)

MARX, J.: Rev. Scien. Instr. **22**, 503 (1951)

MITROCHIN, V. I., REMBEZA, S. I., SPIRIDOV, V. V., JAROSLAVCEV, N. P.: Fiz. Tverd. Tela **27**, 2081 (1985)
NOWICK, A. S., BERRY, B. S.: Anelastic relaxation in crystalline solids; New York, London 1972
PESIN, M. S., POSTNIKOV, V. S., REMBEZA, S. I., JAROSLAVCEV, N. P.: Fiz. Tverd. Tela **18**, 2824 (1976)
SAMUELSON, L.: Physics B **127**, 104 (1984)
SOBOTTA, H.: unpublished research, Karl-Marx-Universität Leipzig 1986
TANAKA, A., SUKEGAWA, T.: J. Appl. Physics **53**, 9208 (1982)
TSUKADA, N., KIKUTA, T., ISHIDA, JK.: Jap. J. Appl. Physics **24**, L689 (1985)
VERNER, V. D., KOVJAZIN, V. Ja., MIL'VIDSKIJ, M. G., OSVENSKIJ, V. S., CHOLODNYI, L. P.: Fiz. Tverd. Tela **19**, 3304 (1977)
VERNER, V. D., NICHUGOVSKII, D. K., KOVYASIN, V. Ya.: phys. stat. sol. (a) **54**, 593 (1979)
WAGNER, G., unpublished research, Karl-Marx-Universität Leipzig 1986
WAGNER, G., GOTTSCHALCH, V., PASEMANN, M.: Acta Phys. Acad. Sc. Hungar. **51**, 391 (1981)

(Received February 2, 1987)

Authors' address:

Dr. D. KLIMM, Prof. P. PAUFLER
Karl-Marx-Universität Leipzig, Sektion Chemie,
Wissenschaftsbereich Kristallographie
Talstraße 35
Leipzig
DDR-7010

Cryst. Res. Technol.	22	1987	8	1031–1039

LECONG DZUONG, H. LÖFFLER, NGUYEN QUYET THANG, HOANG MINH TRI

Hanoi Polytechnical Institute
and Pädagogische Hochschule Halle

The Effect of Manganese Additions on the Structure Changes Proceeding in Rapidly Solidified Al–Cu Alloys

The effect of Mn additions (0 ... 2 wt%) on the decomposition of rapidly solidified Al–4.0 wt% Cu alloys (cooling rate 10^3 to 10^4 K/s; LQ treatment) were studied during ageing between RT and 450 °C by hardness, X-ray methods and electron microscopy. The results were compared with alloys homogenized in the region of the solid solution (SQ treatment).

(i) The LQ treatment results in a quite better homogeneous distribution of the alloyed elements than the SQ one, that is less particles of intermetallic phases are present in the as-cast state. (ii) At $T < 250$ °C Mn additions affect the decomposition kinetics by trapping of vacancies (retardation) and the diminution of the solubility of Cu atoms (acceleration). The first effect dominates in the stage of G.P. zone formation, the second one during precipitation of intermediate phases. (iii) At $T \geqq 300$ °C the intermetallic compound $Cu_2Mn_3Al_{20}$ forms associated with a significant increase of the hardness.

Der Einfluß von Mn-Zugaben (0 ... 2 wt%) auf die Entmischung schnell kristallisierter Al–4.0 wt% Cu-Legierungen (Abkühlgeschwindigkeit von 10^3 bis 10^4 K/s; LQ-Behandlung) wurde mittels Härtemessungen, Röntgenmethoden und Elektronenmikroskopie während der Alterung zwischen RT und 450 °C untersucht. Die Ergebnisse wurden mit Legierungen verglichen, die aus dem Bereich der homogenen festen Lösung abgeschreckt wurden (SQ-Behandlung).

(i) Die LQ-Behandlung ergibt eine wesentlich bessere homogene Verteilung der Legierungselemente im eingeschreckten Zustand als die SQ-Behandlung, d. h. es sind weniger Ausscheidungen intermetallischer Phasen vorhanden. (ii) Bei $T < 250$ °C beeinflussen die Mn-Zugaben die Entmischungskinetik durch Leerstelleneinfang (Verzögerung) und Verminderung der Löslichkeit der Cu-Atome (Beschleunigung). Der erste Effekt dominiert im Bereich der Bildung von GP-Zonen, der zweite bei der Bildung von Zwischenphasen. (iii) Bei $T \gtrless 300$ °C bildet sich die intermetallische $Cu_2Mn_3Al_{20}$-Phase, die eine bedeutende Härteerhöhung bewirkt.

1. Introduction

Among the age-hardenable aluminium alloys, the Al–Cu system belongs to the best investigated ones, as it has been intensively studied for more than six decades. Until the middle of the seventies it was commonly accepted that in the Al–Cu system the decomposition sequence is "Guinier-Preston zones of type I (G. P. zones (I) for short)→ → G.P. zones (II) or $\theta'' \to \theta' \to \theta(Al_2Cu)$". The G.P. zones (I) and the precipitates of the metastable θ''-phase are fully coherent to the ambient aluminium-rich f.c.c. matrix whereas the intermediate θ'-phase has partially lost the full coherency. The tetragonal $\theta(Al_2Cu)$-equilibrium phase is incoherent with respect to the matrix of the aluminium host. As early as 1954 GEROLD proved that the G.P. zones (I), growing in Al–Cu alloys in the room temperature (RT) region, are monolayers of Cu on {100} planes of the matrix. The G.P. zones (II) may consist of two or more Cu layers separated by a few (one to three) Al layers (see e.g. LORIMER, PHILLIPS, 1973, 1975; YOSHIDA et al. 1983; OSAMURA et al. 1982). Recently, however, it was proved (mainly by high-resolution electron microscopy (HREM) and field ion microscopy (FIM)) that the structure changes proceeding in the Al–Cu alloys are more complex than presumed before. Beneath G.P. zones of type I and II, pairs of G.P. zones (I), viz. two Cu layers on {100} matrix planes aligned in parallel and not separated by any Al layer, were found (ABE et al.; YOSHIDA et al., 1976). These multilayer zones have roughened or incomplete surfaces. FIM investigations have shown that G.P. zones (I) may contain holes, however, it was not possible from these experiments to conclude whether these holes correspond to Al atoms or vacancies (OSAMURA et al., 1983; ABE et al.). The formation of pairs of G.P. zones (I) is most probably an intermediate step in the sequence from G.P. zones (I) to (II). A survey about the sequence of phases most likely proceeding in Al–Cu alloys is given in the paper of SCHÜLBE and SCHMIDT.

Rapidly solidified metallic alloys (liquid quench (LQ) techniques) were intensively studied for more than two decades. Among them the Al based alloys are of interest (see, e.g., the review given by ANANTHARAMAN and SURYANARAYANA). The most important advantage of Al based alloys produced by the LQ technique is due to the considerable higher solubility of alloying elements in the liquid state in comparison to the solidified one. Thus, if applying a sufficiently high quenching rate, in the as-cast state of the LQ treatment the portion of the alloyed elements solved in the Al-rich matrix is considerably raised in comparison to the maximum solubility of the solid state according to the equilibrium line of the respective phase diagram. This effect is of particular interest for elements, the solubility of which in the Al lattice is fairly small as, e.g., true for most of the transition elements. For instance for Fe the maximum solubility in the Al lattice amounts only about 0.05 wt%. When applying the LQ technique, the maximum solubility in the metastable solidified state, however, can be increased drastically, in Al–Fe alloys up to about 10 wt%. This supersaturated state of the solid solution can be maintained for fairly long time intervals due to the low mobility of the transition elements in the Al matrix at RT. Thus the LQ technique is of interest from both technological and physical point of view. Among these alloys, rapidly solidified Al–Cu alloys with additions of transition elements have a great perspective for application (increase of the strength, diminution of the stress corrosion sensitivity) and thus have been studied by various authors (see e.g. POLESIA, KOWALENKO 1968, 1970; SUZUKI et al. 1983, 1984). In this paper some results about the effects of additions of Mn on the structure changes proceeding in Al–4.0 wt% Cu–x wt% Mn alloys ($x = 0$; 1.0 and 2.0) are presented. The alloys were solidified with a moderate quenching rate in the order of 10^3 to 10^4 K/s, which can be easily performed even under the conditions of production.

2. Experimental

Samples in the form of small granules were obtained by granulation into water and thin plates by casting in a water-cooled copper mold. The cooling rate just before crystallization was of the order of 10^3 to 10^4 K/s. The obtained microstructure of the solidified specimens was microcrystalline for the granules (mean diameter of the grains in the order of 13 μm) but of columnar type for the plates (mean spacing of about 4 μm). In both cases the samples indicated strong microsegregation of Cu and Mn as proved by microprobe analysis (see DZUONG, THANG). As demonstrated in a previous paper (DZUONG, THANG) the reached cooling rate was sufficient to attain in the as-cast state a supersaturated solid solution of both alloying elements in the lattice of the aluminium host, i.e. no primary intermetallic phases were visible in the electron microscope. This also holds for the alloy with the highest Mn concentration (2.0 wt%) although the maximum solubility of Mn in the Al matrix (binary Al–Mn alloy) only amounts to 1.8 wt% (eutectic point at 657 °C). It can be expected that the additions of Cu rather diminish than raise the solubility of Mn in the lattice of the Al host.

As the effect of fairly large amounts of Mn dissolved in the supersaturated solid solution on the structure changes are intented to be studied, the ageing treatment was performed immediately after solidification, i.e. without any additional homogenization and quenching treatment in the solidified state.

The following experimental methods were applied: microhardness measurements (Vickers hardness with a load of 20 p, HV), precise lattice parameter determination, X-ray phase analysis, small-angle X-ray scattering (SAXS), electron microscopy (TESLA 613 instrument operating at 100 kV) and SEM investigations were made in an instrument of type TSM-35. The composition of the investigated alloys was Al–(3.96 ± 0.06) wt% Cu with Mn additions of 0, (1.08 ± 0.06) and (2.0 ± 0.1) wt%, respectively, and impurities of 0.1 wt% Fe, 0.02 wt% Si and 0.03 wt% Cr.

3. Experimental results and discussion

3.1. Structure changes proceeding during aging at $T_a < 250$ °C

Figure 1 shows the courses of the hardness (HV) during aging at RT, 100 °C and 200 °C. The main features are the following.
(i) In each case, the HV value of the as-cast state is increasing with the rise of the content of Mn.

Considering the curves obtained during aging at RT (Fig. 1a) this increase is $\Delta HV/x_{Mn} = (0.06 \pm 0.01)$ GPa/wt%. According to investigations of BORCHERS et al., the solution hardening effect of Mn in pure Al amounts to $\Delta HV/x_{Mn} = (0.04 \pm \pm 0.01)$ GPa/wt%, when no clustering of Mn atoms occurs. That is a bit less in comparison to the effect of Mn on the HV value of the Al–Cu–Mn alloys (see Fig. 1a). In line with the observed microsegregation of Mn atoms in this case, most probably part of the Mn atoms forms small clusters during the quench to RT, increasing in this manner the hardness of the specimens more efficiently than the single Mn atoms randomly distributed in the matrix.

It is of utmost interest to note that according to investigations of KORNGIEBEL the effect of Mn additions on the hardness of an Al–4.56 wt% Zn–1.14 wt% Mg alloy with $x_{Mn} = 0.85$ wt% is significantly higher, viz. $\Delta HV/x_{Mn} = (0.16 \pm 0.02)$ GPa/wt%, but this alloy was homogenized in the region of solid solution (at 490 °C) and then quenched in water of RT (SQ treatment). The given explanation is that when applying this SQ treatment the tendency of the Mn atoms to form clusters during the quench to RT, and even during the homogenization treatment, is fairly high, since the solubility of Mn in the Al matrix at 490 °C amounts only about 0.4 wt%. The observed differences of the effect of additions of Mn on the hardness between the as-quenched state (SQ treatment) of the Al–Zn–Mg–Mn alloy and the as-cast state (LQ treatment) of the Al–Cu–Mn alloy doubtlessly proves that by applying a LQ treatment even with moderate cooling rates a considerably better homogeneous distribution of the Mn atoms in the matrix can be achieved, that is less particles of intermetallic phases are present in the as-cast state. Obviously, this statement holds for the distribution of the Cu atoms in the matrix, too. According to Figure 1, for the alloy free of Mn, in the as-cast state the HV value is 0.82 GPa. When applying a SQ treatment, however, the respective value is 0.93 GPa (direct quench of an Al–4.3 wt% Cu alloy from 530 °C in oil of the respective aging temperature ranging

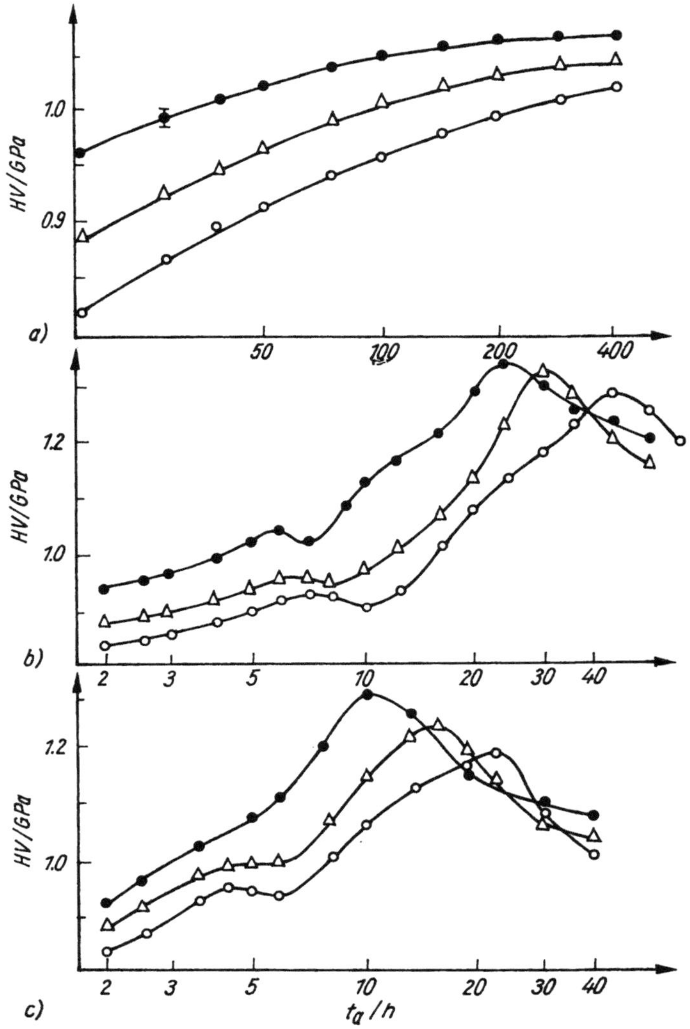

Fig. 1. Course of the microhardness HV during the ageing of the three Al–4 wt% Cu–x wt% Mn alloys (x = 0 (○); x = 1.0 (△), and x = 2.0 (●)) at T_a = RT (a), 100 °C (b), and 200 °C (c), respectively, after the quench from the liquid state (LQ treatment)

between RT and 190 °C, see SCHÜLBE and SCHMIDT; SCHMIDT et al.; SCHÜLBE et al.). Note that $\Delta HV/x_{Cu} = (0.14 \pm 0.02)$ GPa/wt%.

(ii) In the course of the aging at RT the HV curves of the three investigated alloys are approaching more and more, indicating that the favourable effect of the additions of Mn on the hardness is distinctly diminished. This effect was also observed in the Al–Zn–Mg–Mn alloys (KORNGIEBEL).

By SAXS investigations performed with the Al–4 wt% Cu alloys containing 0 and 2.0 wt% Mn, respectively, it could be proved that, actually, in the as-cast state the mean size of the G.P. zones is the same within the range of experimental error, viz.

$r_G = (2.0 \pm 0.2)$ nm, where r_G is Guinier's radius. The size of the G.P. zones of the alloy free of Mn, however, increases faster (2.8 nm after 20 h aging) in comparison to the alloy containing 2 wt% Mn (2.2 nm). Thus one can state that at RT additions of Mn have a retarding effect on the development of the G.P. zones (content of Cu almost 100%). It is most probably due to the trapping effect of vacancies by Mn atoms.

(iii) For all of the applied aging temperatures HV is increasing at least up to 10 h.

In the papers of SCHÜLBE et al. reliable arguments are given that in the alloy free of Mn this increase is due to the formation and growth of precipitates fully coherent to the ambient matrix. That means up to the maximum value of HV the sequence "G.P. zones (I) → pairs of G.P. zones (I) → G.P. zones (II)" is proceeding. Applying $T_a = 100$ °C and 200 °C a short plateau, or even a faint maximum, of HV is observed at about 0.95 GPa (see Fig. 1). Then the hardness is once more increasing, reaching an absolute maximum level of about 1.25 GPa.

Although the Cu content of the alloy investigated by SCHÜLBE and SCHMIDT (SQ-treatment) is about 0.3 wt% higher than the alloys investigated in the present paper (LQ treatment), striking differences of the courses of HV are visible (see Fig. 2).
(iv) When applying the SQ treatment, during the aging at 100 °C no maximum of the hardness (HV_{max}) is observed (overaging effect), but it is distinctly visible in the LQ samples.

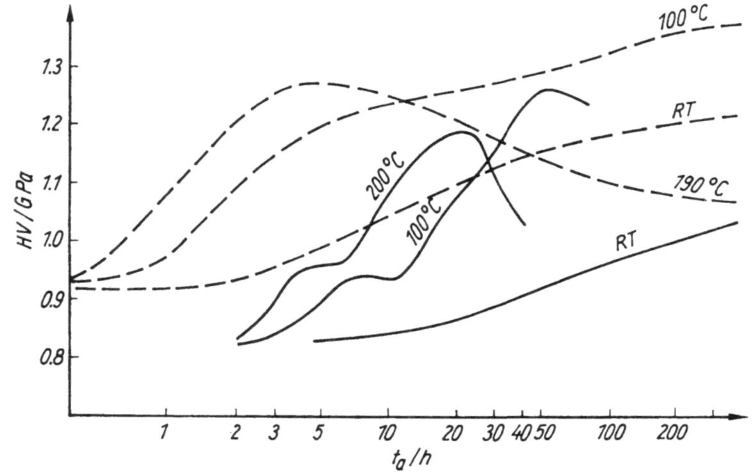

Fig. 2. Comparison of the courses of HV of the alloys free of manganese during the ageing at various T_a between the quenching treatments: LQ (Al–4 wt% Cu) (———); SQ (Al–4.3 wt% Cu) (– – –)

It is generally accepted that in Al–Cu alloys the overaging effect is due to the transition of G.P. zones (II) into particles of the metastable θ'-phase (break down of the full coherency) and that this transition is accelerated with increasing Cu content. In our opinion, in the Al–Cu alloy exposed to the LQ treatment, this transition is favoured by the lattice faults present in a high density in the as-cast state. They increase the strains in the matrix considerably. It is known that lattice strains diminish the critical size of metastable precipitates to be transformed into more stable phases (see, e.g., LORIMER 1979; LÖFFLER et al.).

The occurrence of this transition is supported by electron microscopic investigations. The micrographs taken from the specimens showing peak hardness ($T_a = 100$ and 200 °C) display the presence of G.P. zones (II) and precipitates of the θ'-phase, the density of which was higher in the regions close to dislocations, grain and subgrain boundaries.
(v) During the aging at elevated T_a, viz. 100 and 200 °C, a maximum of HV is observed. The position of the peak hardness is shifted to shorter ageing times the higher the content of Mn is. Obviously additions of Mn have two main effects on the kinetics of the structure changes, namely a retarding effect, it is due to the trapping of vacancies, and an accelerating one, namely the diminution of the solubility of Cu atoms in the matrix. The first effect dominates in the stage of G.P. zone formation, the second one during precipitation of intermediate phases.

3.2. Aging at $T \geqq 300$ °C

In Figure 3 the courses of HV during the aging at T_a ranging between 300 and 450 °C are presented for the alloy free of Mn (bottom part) as well as for the alloy with 2 wt% Mn (upper part). Plates produced by the LQ treatment were investigated. The specimens were preaged at RT for about 30 d.

When exposing the alloy free of Mn to $T_a > 250$ °C only a monotonous diminution of HV was observed. Thus in Figure 3 for this alloy only one curve is drawn (400 °C). The reason of this monotonous decrease is that the dissolution line of G.P. zones (I), grown in an Al–4.0 wt% Cu alloy at RT, is situated not higher that 210 °C (see LORIMER, NICHOLSON; FOUQUET et al.; BORELIUS et al.) and even the dissolution temperature of G. P. zones (II) is placed not higher than about 270 °C (HIRANO, IWASAKI; ZAHRA, LAFFITTE; ZAHRA et al.). The most striking features of the hardness curves obtained for the Al–4 wt% Cu–2.0 wt% Mn alloy are the following:
(i) As has to be expected, the first step of the structure changes proceeding during the ageing at $T_a \geqq 300$ °C is the dissolution of G.P. zones (I) grown during the preaging at RT. This is indicated by the diminution of HV.

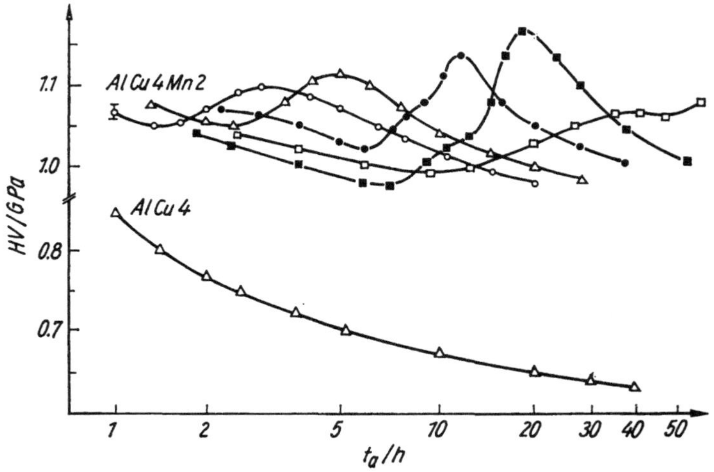

Fig. 3. Course of the microhardness HV during ageing of plate-shaped specimens at elevated T_a; upper part: AlCu4Mn2 alloy; T_a = 300 (□), 350 (■), 375 (●), 400 (△), 450 (○) °C. AlCu4 alloy; T_a = 400 °C

(ii) After passing a minimum the hardness is increasing up to a maximum and then decreasing again (overaging effect). As has to be expected, the positions of both minima and maxima are shifted to shorter aging times with increasing T_a.

As proved by electron- as well as X-ray diffraction pattern this increase of HV is not due to the formation of Al–Cu phases (θ' and θ), but to particles of the intermetallic compound $Cu_2Mn_3Al_{20}$ unevenly distributed in the grains.

Figure 4 shows, as an example, the structure of the alloy (granule) after aging at 400 °C for 8 h (that is short after peak hardness, see Fig. 5). The precipitates have a spherical or ellipsoidal shape and a mean diameter of about 170 nm. In line with the data reported in the literature, according to our diffraction pattern the crystal structure of the particles of the $Cu_2Mn_3Al_{20}$-(T-)phase is orthorhombic with the lattice parameters $a = (2.45 \pm 0.01)$ nm, $b = (1.28 \pm 0.01)$ nm and $c = (0.14 \pm 0.01)$ nm. The particles of the T-phase preferentially form at dislocations and subgrain boundaries (see Fig. 4) where according to microprobe analysis the segregation of Mn atoms is fairly high.

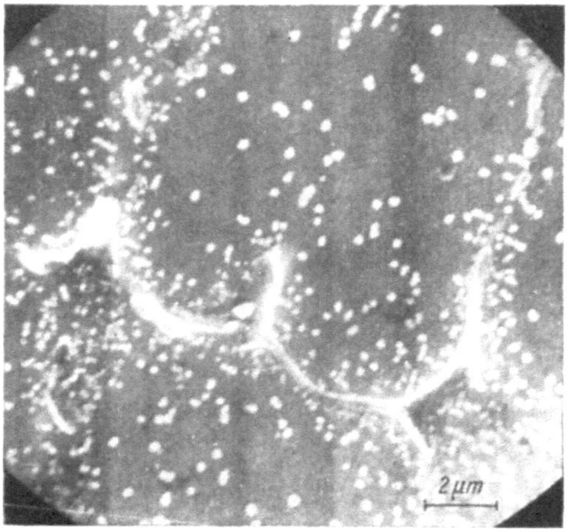

Fig. 4. Structure of the AlCu4Mn2 alloy (granule) after ageing at 400 °C for 8 h (after the peak hardness, see Fig. 5) SEM, operating at 25 kV

(iii) With the formation of the T-phase another interesting feature was observed. From the X-ray diffraction pattern it could be concluded doubtlessly that the formation of this phase is associated with a splitting of the lattice parameter a_m of the aluminium-rich matrix (see Fig. 5). This effect can be explained on the basis of the following conception. Both the atomic radii of the Cu atoms ($r_{Cu} = 0.127$ nm) and Mn atoms ($r_{Mn} = 0.130$ nm) are about 10% smaller than the Al atoms ($r_{Al} = 0.143$ nm). As the consequence, the dissolution of the G.P. zones (I) formed during the preageing at RT, occurring until the minimum of HV is reached at $T_a = 300$ °C, results in a decrease of the lattice constant a_m. The diminution of the concentrations of Mn and Cu atoms in the matrix, however, leads to an increase of a_m. As shown in Figure 5, during the diminution of HV in the first period of the ageing at 400 °C, a_m is not decreasing, but is remaining at first constant and is then slightly increasing. This behaviour demonstrates that the dissolution of the G.P. zones (I) (increase of a_m) and the formation of precipitates of the T-phase are overlapping processes. The splitting of a_m occurs when precipitates of the T-phase, sufficiently large in size

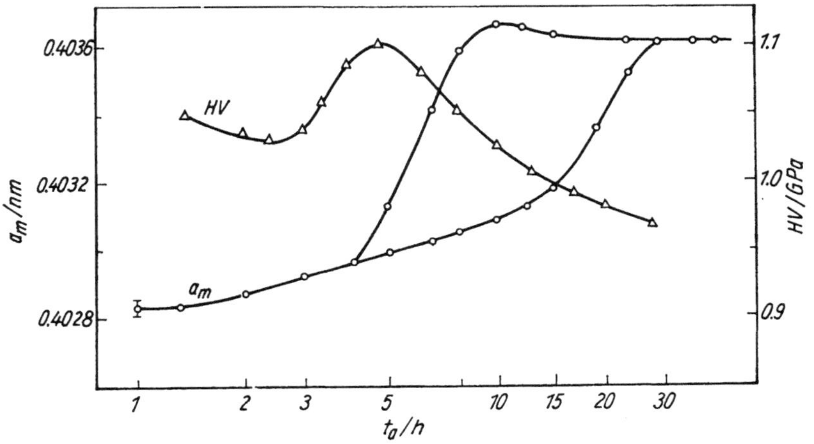

Fig. 5. Courses of the microhardness HV and lattice parameter a_m of the Al-rich matrix during the ageing of granules of the AlCu4Mn2 alloy at T_a = 400 °C. a_m was determined from X-ray diffraction pattern, CuK_α radiation

start to form. As proved by Figure 4 these precipitates are unevenly distributed in the grains. In the regions with a high density of particles of the T-phase the matrix is considerably more exhausted on Cu and Mn atoms than in the regions were only few precipitates are present. As indicated by the strong decrease of the hardness after reaching peak hardness the particles of the T-phase start to coarsen and the distribution of the precipitates becomes more and more even over the whole grain, resulting in a uniform value of a_m (a_m = 0.4036 nm).

4. Conclusions

(i) Even when solidifying the investigated alloys with the rather moderate cooling rate of 10^3 to 10^4 K/s, in the as-cast state, a considerably better homogeneous distribution of the alloyed elements can be achieved in comparison to the SQ treatment, that is less particles of intermetallic phases are present.

(ii) At $T_a < 250$ °C the additions of Mn affect the kinetics of the structure changes mainly by two effects, viz. by trapping of free vacancies (retarding effect) and by the diminution of the solubility of Cu atoms in the Al-rich matrix (accelerating effect). The first effect dominates in the stage of G.P. zone formation, the second one during precipitation of intermediate phases.

(iii) At $T_a \gtrapprox 300$ °C precipitates of the intermetallic compound $Cu_2Mn_3Al_{20}$ (T-phase) form. Its formation is associated with a significant increase of the hardness even at $T_a = 450$ °C. It indicates that, at least up to peak hardness, the precipitates are fairly finely distributed in the matrix and that they are most probably surrounded by stresses. Specimens of the Al–4.0 wt% Cu–2.0 wt% Mn alloy aged at, e.g., $T_a =$ = 350 °C to peak hardness may result in good temperature resistance.

(iv) The formation of the T-phase is accompanied by splitting of the lattice constant a_m of the Al-rich matrix. The reason is that in the vicinity of dislocations and subgrain boundaries the density of the precipitates of the T-phase is considerably higher as compared with the regions free of these lattice faults.

References

ABE, T., MIYAZAKI, K., HIRANO, K.: Acta metallurg. **30**, 357 (1982)

ANANTHARAMAN, T. R., C. SURYANARAYANA, I.: Mater. Sci. **6**, 1111 (1971)

BORCHERS, G., ANDERSSON, J., GULLBERG, K.: Ing. Vetenskaps Adad. Hanoll, n° 169 (1943)

BROCHERS, H., KNOBLOCH, H., THYM, G.: Metall **25**, 225 (1971)

DZUONG, L., THANG, N. Q.: in: Proc. 4th Intern. Conf. Age-hard. Al Alloys, Balatonfüred, Hungary, 1986, P36

DZUONG, L., KHOA, T. S.: hoc Kythuat Luyenkim, DHBK, 1982, p. 29

FOUQUET, F., COHAN, M.; MERLIN, J., FANTOZZI, G., GOBIN, G. F., DUVILLARD-CASKE, C., WINTERBERGER, M.: Mem. Scr. Rev. Met. **70**, 683 (1973)

GEROLD, V.: Z. Metallkde. **45**, 593 (1954)

HIRANO, K., IWASAKI, H.: Trans. Jap. Inst. Metals **5**, 162 (1964)

KORNGIEBEL, E.: Neue Hütte **29**, 426 (1984)

LÖFFLER, H., SYNĚCEK, V., SIMERSKA, M., WENDROCK, G., BARTUŠKA, P., KROGGEL, R.: phys. stat. sol. (a) **65**, 197 (1981)

LORIMER, G. W., NICHOLSEN, R. B.: in: The Mechanism of Phase Transformations in Crystalline Solids, Institute of Metals Monograph + Raport Series **33**, 36 (1969)

LORIMER, G. W.: in: Proc. Phase Transf. Spring Resid. Conf., London **2**, 69 (1970)

LORIMER, G. W.: in: Solid-solid Phase Transformation (ed. by H. I. AARONSON), The Metall. Soc. of AIME, New York 1982, p. 613

OSAMURA, K., OTSUKA, N., MURAKAMI, Y.: Phil. Mag. **B45**, 583 (1982)

OSAMURA, K., MURAKAMI, Y., SATO, T., TAKAHASKI, T., ABE, T., HIRANO, K.: Acta metallurg. **31**, 1669 (1983)

PHILLIPS, V. A.: Acta metallurg. **21**, 219 (1973)

PHILLIPS, V. A.: Acta metallurg. **23**, 751 (1975)

POLESIA, A. F., KOWALENKO, B. B.: Fiz. met. **27**, 479 (1968)

POLESIA, A. F., KOWALENKO, B. B.: Izv. Acad. Nauk SSSR, Met. **1**, 173 (1970)

SCHMIDT, U., ZAHRA, A., SCHÜLBE, R., LÖFFLER, H.: phys. stat. sol. (a) **95**, 73 (1986)

SCHÜLBE, R., SCHMIDT, U.: phys. stat. sol. (a) **91**, 433 (1985)

SCHÜLBE, R., SIEBERT, P., SCHMIDT, U.: phys. stat. sol. (a)

SUZUKI, H., ITOH, G., KOYAMA, K.: J. Jap. Inst. Metals **47**, 873 (1983)

SUZUKI, H., ITOH, G., KOYAMA, K.: J. Jap. Inst. Light Metals **34**, 227 (1984)

YOSHIDA, H., COCKAYNE, D. J. H., WHELAN, M. J.: Phil. Mag. **34**, 89 (1976)

YOSHIDA, H., HASHIMOTO, H., YOKOTA, Y., AJIKA, N.: Trans. Jap. Inst. Metals **34**, 378 (1983)

ZAHRA, A., LAFFITTE, M.: Scripta metall. **8**, 165 (1974)

ZAHRA, A., ZAHRA, C. Y., LAFFITTE, M.: Scripta metall. **9**, 879 (1975)

(Received February 2, 1987)

Author's addressesn:

Dr. LECONG DZUONG (now in Halle (Saale)), NGUYEN QUYET THANG, HOANG MINH TRI
Hanoi Polytechnical Institute, Hanoi, Vietnam,
Truong DHBK Hanoi, Luyenkim

Prof. Dr. habil. H. LÖFFLER
Pädagogische Hochschule Halle, Sektion Mathematik/Physik
Kröllwitzer Str. 44
Halle (Saale)
DDR-4050

Cryst. Res. Technol.	22	1987	8	1041–1046

F. PECHAR

Czechoslovak Academy of Sciences, Institute of Geology and Geotechnics,
Prague, Czechoslovakia

An X-ray Diffraction Refinement of the Crystal Structure of Natural Apophyllite

The crystal structure of apophyllite (Andersberg — GDR), with $a = 8.966(2)$, $c = 15.767(1)$ Å, P4/mnc, $z = 2$, $\varrho = 2.35$ gm · cm^{-3}, has been refined (R = 0.035) by least squares with 976 reflections collected with a diffractometer. Apophyllite is a sheet structure with the bringing Si–O bond lengths 1.6236 Å, while the non-bridging bond length Si–O(3) is 1.5843 Å. The two independent Si–O–Si angles are 140.09, 140.76 degrees. The mean Si–O bond length is 1.6138 Å.

As it was not possible to locate unequivocally the H-atoms with the X-ray data, the interpretation of the of water in the structure was based on the charge balance approach of DONNAY and ALLMAN. Assuming that the fluorine ion is bonded to calcium ion (Ca–F = 2.416 Å) and to H3 atom (F–H3 = 0.9145(1) Å) and water molecule hydrogen bonded to silicate framework. The average bond distances O–H are 0.962(2) Å and angle 105.24(1)°.

Introduction

Apophyllite is an unusual sheet structure with space group P4/mnc (TAYLOR, NÁRAY-SZABÓ). Rings of four silicon-oxygen tetrahedra surround the four-fold axes and form parts of rings with eight tetrahedra resulting in a layered structure. The unshared oxygens in the adjacent four-fold rings point approximately along the $+c$ and $-c$ crystallographic axes, respectively. The coordination of the interlayer ions is also unusual; the potassium ion is surrounded by eight water molecules, the calcium ion is bound to two water molecules (shared with K^+) and four oxygen ions, forming a trigonal prism. The fluorine ion has been considered to be part of the Ca-polyhedron (COLVILLE et al.; PRINCE); however, PRINCE 1970, hereafter referred to as Part II finds that fluorine is part of an HF group. The fluorine ion is surrounded by a planar group of four Ca ions.

A number of interesting aspects of the structure and physical properties prompted us to embark on a refinement of the crystal structure. We were curions about the precise average bond length that would be found in a four-unit sheet structure with pure silicon oxygen tetrahedra. Gillespite (PABST) and isostructural compounds such as "Egyptian blue" (PABST) have a comparable silicate sheet. In these structures, there is more overlap in adjacent four-fold rings resulting in a smaller value for the a cell edge.

In addition, a recent refinement of a related structure, datolite (PANT and CRUICKSHANK) showed large variations in the silicon-oxygen bond distances within a single tetrahedron, with the nonbridging Si–O distances significantly shorter than the bridging distances. Apophyllite has a similar structure and might be expected to show similar large variations in bond lengths, particularly since the interlayer Ca^{2+} ions have low electronegativity. The position of the protons belonging to the water molecule and their effect on the physical properties have been a problem for sometime (COLVILLE et al.; CHAO; PRINCE).

In the original determination of the structure of apophyllite (TAYLOR, NÁRAY-SZABÓ), the data were adequate to note one short interatomic distance between the oxygen of water molecule and another oxygen in the silicon tetrahedron. A consideration of bond strengths led them to the conclusion that the water oxygen position is occupied by an OH — which is linked to O(3) by a proton.

Experimental

Small enhedral crystals (0.3 mm) from Andersberg (GDR), were used in this study. Morphological analysis as well as Weissenberg and precission photographs taken of these crystals indicate no deviation from the preivously reported space group P4/mnc (TAYLOR, NÁRAY-SZABÓ). The cell dimensions derived from single crystal work were refined by least squares method using powder diffraction data (CuKα radiation, α-Al_2O_3 as internal standard) shows Table 1 with comparison to previous publications.

Table 1

The cell constants of apophyllite with comparison to previous publication

cell constants	this work	COLVILLE 1971	CHAO 1971
a/Å	8.966(2)	8.965(2)	8.965(3)
b/Å	8.966(2)	8.965(2)	8.965(3)
c/Å	15.767(1)	15.768(2)	15.767(7)

Its chemical composition expresses the crystallochemical formula $(Na_{0.12}K_{0.88})Ca_4 \cdot \cdot (Si_8O_{20})F \cdot 8\ H_2O$.

The diffraction data were collected on the automatic four-circle X-ray diffractometer Syntex $P\bar{1}$ (ÚMC HČSAV Praha, Dr. Hašek). The wavelength of the MoKα radiation was 0.71073 Å. The intensity data were collected by intervals $\sin\theta/\lambda = 0.60$.

Altogether 976 reflections were measured in the MoKα sphere.

The intensities were corrected for Lorentz and polarization factors. Absorption corrections were not significant since $\mu = 0.15\ cm^{-1}$. The refinement was carried out with a least-squares program (CRYLSQ; XTAL system) on the Siemens computer. During the first cycle of refinement, the atomic positions of COLVILLE et al. were used. Form factors were taken from International Tables, Vol. III. Varying the atomic parameter and one scale factor an holding the isotropic temperature factors constant led to a residual R of 0.20.

The final anisotropic refinement gave $R = 0.035$. Table 2 shows the final atomic positions and the anisotropic temperature factors. Table 3 shows bond distances and angles (Bondla program in the XTAL system). Figure 1 shows the structural motive of the environment of the water molecule. Figure 2 shows the Si_8CO_{20} sheet in apophyllite projected on (001) and Figure 3 shows the K, Ca sheet projected on (001).

Discussion

The structure of apophyllite is shown in Figure 1÷3. In the tetrahedral layers, only the unshared oxygen 03, is bonded to the interlayer Ca ions (Tab. 3). It can be seen that the oxygen O4, of the water molecule occupies a strategic position in the structure. The Ca^{2+} ion is coordinated by eight O4 atoms in the form of a square tetragonal prism. Each O4 is shared with a Ca^{2+} ion which, in turn, includes two O4 atoms among

Table 2

Final atomic fractional coordinates, thermal parameters ($10^6 \times$)

Atom	occupancy factor	x	y	z
Ca	0.98(1)	0.1096(1)	0.2462(1)	0
Si	1.91(1)	0.2256(1)	0.0865(1)	0.1900(1)
O1	1.01(1)	0.3631(3)	0.1369(3)	0.25
O2	1.00(1)	0.846(2)	0.1891(3)	0.2178(2)
O3	1.01(1)	0.2636(3)	0.1026(3)	0.0923(2)
O4	0.95(2)	0.2131(3)	0.4491(3)	0.0898(2)
K	0.92(3)	0	0	0.5
F	0.97(3)	0	0	0
H1	0.95	0.4490(4)	0.1730(4)	0.0889(2)
H2	0.95	0.2246(4)	0.4255(4)	0.1477(2)
H3	0.25	0	0	0.0580(10)

Atom	B11	B22	B33	B12	B13	B23
Ca	2773(98)	3202(98)	967(30)	75(15)	9	0
Si	2285(98)	2039(98)	959(30)	33(15)	− 59(10)	− 11(10)
O1	3936(400)	3936(400)	1327(100)	−631(600)	−275(500)	275(512)
O2	3412(200)	4789(200)	1965(90)	1246(400)	−653(270)	−617(200)
O3	5001(300)	4314(200)	1135(70)	55(70)	−582(200)	−307(200)
O4	8905(400)	4494(300)	1449(90)	88(80)	−202(300)	−670(300)
K	7615(400)	7615(400)	4376(150)	0	0	0
F	2674(700)	2674(700)	2580(250)	0	0	0
H1	2000(300)	2600(500)	2600(500)	−100(100)	360(90)	− 60(50)
H2	5000(600)	2600(500)	1300(500)	0	−500(100)	100(100)
H3	7100(800)	7100(800)	2800(900)	0	0	0

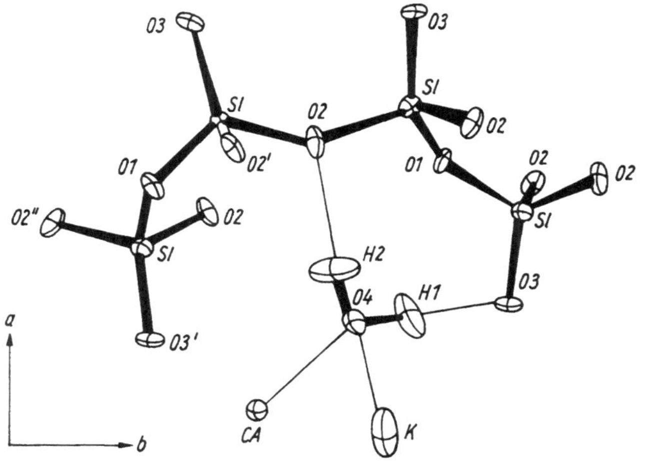

Fig. 1. The ORTEP plot of the structural motive of apophyllite and the environment of the water molecule

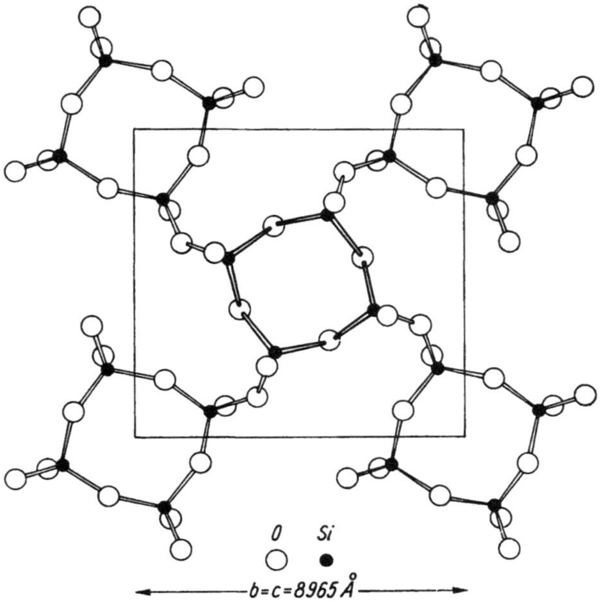

Fig. 2. The Si_8O_{20} sheet in apophyllite structure projected on (001)

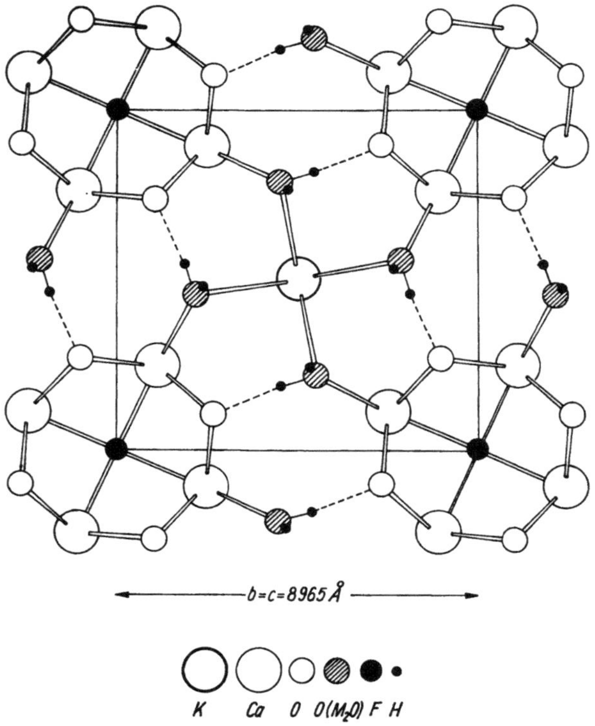

Fig. 3. The K, Ca sheet in apophyllite projected on (001)

Table 3

Selected interatomic distances (Å) and angles (°) in apophyllite (E.S.D. in parentheses)

Bond	distance	bond	angle
Si–O1	1.6183(2)	O1-Si-O2	106.09(1)
Si–O2	1.6238(2)	O1-Si-O3	112.28(1)
Si–O3	1.5843(2)	O1-Si-O2^{i}	104.99(1)
Si–O2^{i}	1.6287(3)	O2–Si–O3	112.23(1)
		O2–Si–O2	107.73(1)
		O3–Si–O2^{i}	112.99(1)
		Si–O1–Si	140.10(1)
		Si–O2–Si	140.76(1)
Ca–O3	2.3931(3)2 ×	O3–Ca–O3	96.39(1)2 ×
Ca–O3^{i}	2.3905(4)2 ×	O3–Ca–O3	75.38(1)2 ×
		O3–Ca–O3	149.29(1)2 ×
Ca–F	2.4163(5)4 ×	O3–Ca–F	74.62(1)
K–O4	2.975(3)8 ×		
F–H3	0.9145(1)	Ca–F–H3	90.0
		Ca–F–Ca	180.0
		H3–F–H3	180.0
H1–O4	0.9819(2)	H1–O4–H2	105.24(1)
H2–O4	0.9248(1)	O3–H1–O4	161.0(2)
H1–O3	1.8272(1)	O2–H2–O4	147.7(2)
H2–O2	2.6933(2)		

its six nearest neighbours forming a trigonal prism. Theoretical studies of bond length and bond angle variations in silicates (BROWN, GIBBS) can be tested against apophyllite data. As predict by PAULING an later authors, the bridging Si–O bonds (mean length = = 1.625 Å) should be, and are, longer than the nonbridging Si-O3 bonds of 1.5843(2)Å. If appreciable π bonding is responsible for the short bond, the bond angles subtended at silicon by the tetrahedral edges and involving O3 should be affected.

They are indeed, with angles 112° to 113°, whereas the angles subtended the method of DONNAY and ALLMAN. It is instructive to calculate the valence using the model that would have been assumed from X-ray data alone, namely, that the fluorine ion part is of the Ca-polyhedron.

Figure 1 shows the water molecule and its environment. H1 clearly forms a strong hydrogen bond between O3 and O4. The O4–H2 bond points toward O2, with an O4–H2–O2 bond angle of 147.7°. HAMILTON and IBERS suggest that an A-H-B linkage can be considered a hydrogen bond, its distance H-B is less than the distance determined by subtracting 0.2 Å from the sum of the van der Waals radii. It the B atom is oxygen, this distance is about 2.4 Å, so O4–H2–O2 can be considered a weak hydrogen bond.

The H1–O4 distance is longer than the H2–O4 distance, which is consistent with the fact that H1 is involved in a stronger hydrogen bond.

The fluorine atom is coordinated to four Ca atoms (2.4163 Å) as well as to the H3 atom (0.9145 Å). The valence summation used to predict the proton positions assumes that fluorine is present as the F^- ion, and that it participates in the calcium coordination polyhedron. However, if flourine is bonded to hydrogen in HF molecules, the calcium flourine interaction is weakened, and the proper valence summation around the Ca ion is quite uncertain. A limiting case can be represented by ignoring the Ca–F interaction, and making the coordination number for Ca = 6 rather than 7. Since the H position are known, hydrogen is included as a cation, and its valence contributions have been estimated using the DONNAY and ALLMAN formula for the long bond valence, v, and taking the short bond valence v_s, equal to $1 - v$. This procedure gives lightly too much charge to O3, and not quite enough to O4, but the agreement is remarkably good. If a large number of similar structures were known, it might the possible to construct an empirical relation connecting the bond valences with the O–H distances (PRINCE).

Summary

A refinement of the structure of apophyllite has given precise values for the variation of Si–O bond lengths and for the mean Si–O distance in this sheet structure. It has also provided data on the coordination geometry of Ca^{2+} and K. O4 is found to belong to the water molecule, the positions of hydrogen bonds between oxygen ions and also the H–F bond have been determined by applying the principle of local charge neutralization in a quantitative way.

References

BROWN, G. E., GIBBS, G. V.: Amer. Mineralog. **54**, 1528 (1969)
CHAO, G. Y.: Amer. Mineralog. **56**, 1234 (1971)
COLVILLE, A. A., ANDERSON, C. P., BLACK, P.: Amer. Mineralog. **55**, 1220 (1971)
DONNAY, G., ALLMAN, R.: Amer. Mineralog. **55**, 1003 (1970)
HAMILTON, W. C., IBERS, J. A.: Hydrogen Bonding in Solids, New York 1968
PABST, A.: Acta Cryst. **12**, 733 (1959)
PANT, A. K., CRUICKSHANK, D. W. K.: Z. Krist. **125**, 186 (1967)
PAULING, L.: J. Amer. Chem. Soc. **51**, 1010 (1929)
PRINCE, E.: Amer. Mineralog. **56**, 1243 (1971)
STEWART, J. M., HALL, S. R. (eds): The XTAL System of Crystallographic Programs, Computer Sci. Techn. Rep. Ser. University of Maryland 20742, 1983
TAYLOR, W. H., NÁRAY-SZABÓ, St.: Z. Krist. **77**, 146 (1931)

(Received February 2, 1987)

Author's address:

Dr. F. PECHAR
Czechoslovak Academy of Sciences, Institute of Geology and Geotechnics
V Holesovickach 41
Prague 8, CS-182 09
Czechoslovakia

Cryst. Res. Technol.	22	1987	8	1047–1054

T. PALEWSKI, J. HORN

International Laboratory of High Magnetic Fields and Low Temperatures, Wrocław, Poland, and Institute of Inorganic Chemistry and Metallurgy of Rare Elements, Technical University, Wrocław, Poland

Magnetically Induced Rhombohedral Distortion in the $U_xTh_{1-x}Se$ and $U_xY_{1-x}Se$ Solid Solutions

Solid solutions of $U_xTh_{1-x}Se$ for $x = 1.00 \div 0.80$ and the solid solutions of $U_xY_{1-x}Se$ for $x = 1.00 \div 0.60$ were studied by the X-ray diffraction over the temperature range $100 \div 200$ K. In both systems rhombohedral distortions of the cubic rock-salt crystal structure was observed below the Curie temperatures. The temperature and composition dependences of the rhombohedral angle ω of the rhomboedral cell and the $\frac{c' - a'}{a'}$ ratio were determined.

Были сделаны рентгеновские исследования твердых растворов $U_xTh_{1-x}Se$, для которых $x = 1,00 \div 0,80$ и твердых растворов $U_xY_{1-x}Se$, для которых $x = 1,00 \div \div 0,60$. Измерения велись в температурной области $100 \div 200$ K. Ниже температуры ферромагнитного упорядочения (T_c) обнаружено, что кубическая кристаллическая структура типа хлорида натрия имеет ромбоэдрическое искажение. Была определена зависимость угла ромбоэдрической ячейки (ω) и отношения $\frac{c' - a'}{a'}$ от температуры и состава исследованных твердых растворов.

1. Introduction

The monoselenides of uranium, thorium and yttrium have the rock-salt crystal structures, their values of the lattice constants being close to each other. The lattice constant of USe was found to range between $a = 0.574$ nm and 0.576 nm (FREEMAN, DARBY) and most probably its real chemical formula is USe_x where $x \leqq 1$. This suggestion is also supported by a spread of experimentally determined Curie temperatures. USe was found to be ferromagnetic with the Curie temperatures ranging between 161 K and 182 K (TRZEBIATOWSKI, SUSKI; TAKAHASHI, WESTRUM). Neutron diffraction studies showed that the direction of the magnetic moments of uranium in USe is along [111] axis (VOGT). WEDGWOOD and KUZNIETZ suggested that ferromagnetic exchange interactions may change the symmetry of the crystal lattice of uranium chalcogenides from cubic to rhombohedral (WEDGWOOD, KUZNIETZ). This suggestion was then confirmed by MARPLES in his low temperature X-ray studies of US and USe (MARPLES; MARPLES et al.). These studies have revealed that for USe the rhombohedral angle ω decreases below the Curie temperature and reaches at 4.2 K a value of about 0.3° less than 90°. ALDRED et al. found rhombohedral distortion to occur also in UTe. ALLBUTT et al. has performed a similar studies on the US-UP solid solution and confirmed the occurrence of the distortion of the crystal cell in ferromagnetic solid solutions US-UP below T_c. It was found that distortion takes place in the ferromagnetic region of the USe-UAs solid solutions (PALEWSKI et al.

1970). It is clear that the reason of this rhombohedral distortion are the ferromagnetic exchange interactions existing in these compounds and their solid solutions. From this point of view it was interesting to examine how a weakening of the ferromagnetic interactions influenced this distortion. One of obvious ways to make the exchange interaction weak is diluting the ferromagnetic substance with a diamagnetic one. In the case of USe such an isostructural but diamagnetic substance may be ThSe, LaSe, or YSe, if their lattice parameters are close to that of USe. It is known from the literature (D'EYE et al.; JANDELLI) that the lattice constant of ThSe is equal 0.5863 nm and that of YSe amounts to 0.5703 nm. These values of the lattice constants suggested that it would be possible to prepare solid solutions of USe with ThSe or YSe.

2. Experimental

2.1. Sample preparation

The $U_xTh_{1-x}Se$ and $U_xY_{1-x}Se$ solid solutions were prepared from mixtures of the selenides, USe and ThSe or USe and YSe, by melting their cold-pressed tablets in an arc furnace under the purified argon. The buttons of the solid solutions were annealed at 800—900 °C for 2÷4 days.

The monocompounds USe, ThSe and YSe were prepared by heating the cold-pressed mixtures of the pulverized elements in evacuated silica tubes with tantalum crucibles at 1000 °C during one week. After this preliminary synthesis, the products were crushed to form fine powders, pressed again and heated in an induction furnace under vacuum at the temperature 1660—1700 °C for one hour. The quality of the monoselenides and their solid solutions was controlled by X-ray diffraction and the magnetic measurements. The X-ray patterns indicated NaCl-type crystal structures for the pure monoselenides and most compositions of the solid solutions, and only $U_xY_{1-x}Se$ samples with x close to 0.60 exhibited a few very weak lines belonging to another phase. The magnetic measurements were carried out by the Faraday method over the temperature range 4.2 K to 300 K at the external field below 1 T employing a Cahn electrobalance. These studies confirmed the existence in broad region of composition of the ferromagnetic properties of solid solutions and diamagnetic ones of ThSe and YSe.

2.2. Details of experiment

The low temperature X-ray experiments were carried out by using a "DRON" diffractometer with CuK_α radiation. The powdered samples of USe and its solid solutions with ThSe or YSe used in the X-ray studies were mixed with an organic oil and put into a copper container joined to the diffractometer axis. The samples together with the container were surrounded by a thermally-isolated box. The samples were cooled with cold nitrogen vapours flown through the box. The system used in our set up allowed us to curry out the measurements at any temperature between 100 and 300 K. The stability of the temperature was established to be 1÷2 K. A chart recorder system recorded the profiles of the rocking curves.

3. Results and discussion

3.1. Magnetic measurements

It has been established that the distortion of crystal lattice in actinide compounds is caused by magnetic order existing in these compounds (LANDER, MUELLER). The same seemed to be true in $U_xA_{1-x}Se$ solid solutions (A = Th or Y), thus, it was necessary to perform magnetic measurements before carrying out structural studies.

It was found that the ferromagnetic region for the $U_xTh_{1-x}Se$ system extend from $x = 1.00$ to $x = 0.50$ (PALEWSKI 1984). In the case of the $U_xY_{1-x}Se$ solid solutions were prepared with x changing from 1.00 to 0.60. All compositions of the latter system were found to possess ferromagnetic ordering below 160 K. The composition dependences of the lattice constants and the Curie temperature for $U_xTh_{1-x}Se$ and $U_xY_{1-x}Se$ systems are presented in Figure 1. As is seen from Figure 1a, for $U_xTh_{1-x}Se$ system the dependence of the lattice parameter exhibits a slight negative deviation from the linear course over the whole range of compositions. The dependence of T_c on the composition is almost linear except for the region close to $x = 1.00$. The picture is different in the $U_xY_{1-x}Se$ solid solutions (see Fig. 1b). The lattice constants in the system are not much different from the parameter of USe. The dependence of the Curie temperature on the composition is rather weak except for a flat maximum with $0.95 \geqq x \geqq 0.90$. The values of T_c for samples with $0.85 \geqq x \geqq 0.60$ are constant ($157 \div 158$ K).

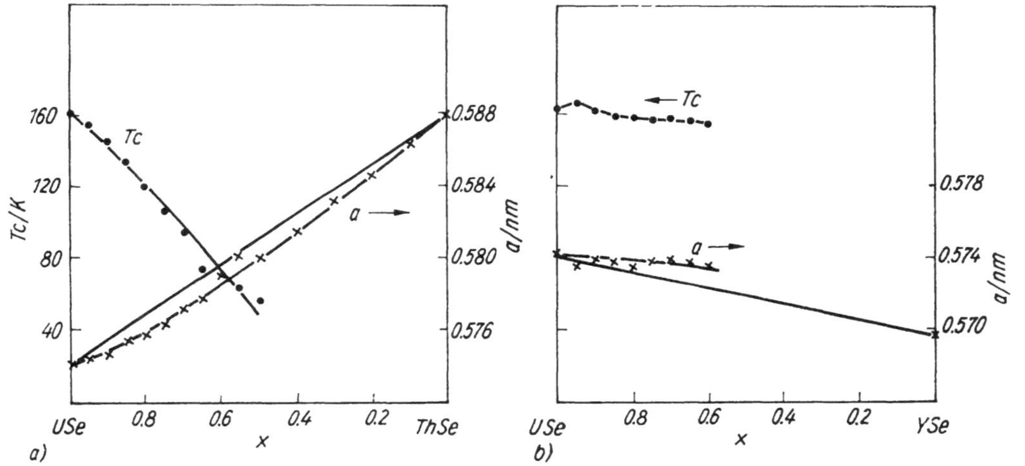

Fig. 1. a) Structural and magnetic data of $U_xTh_{1-x}Se$ system; b) Structural and magnetic data of $U_xY_{1-x}Se$ system

3.2 Rhombohedral distortion

The set of the crystallographic planes of the cubic lattice with the Miller indices hkl, $\bar{h}kl$, $\bar{h}\bar{k}l$, and $\bar{h}\bar{k}\bar{l}$ gives the reflections at the same Bragg angle. However, after the distortion of the cubic cell along the [111] axis to the rhombohedral one, the reflection due to hkl, $\bar{h}kl$, and $\bar{h}\bar{k}l$ crystallographic planes appear at different Bragg angles. The relation between the rhombohedral angle (ω) of the distorted cubic cell with ω near 90° and the position of the reflection (θ_{hkl}) may be expressed as:

$$\frac{4 \sin^2 \theta_{hkl}}{n^2 \cdot \lambda^2} = \frac{(h^2 + k^2 + l^2) \cdot \sin^2 \omega + 2(hk + kl + lh) \cdot (\cos^2 \omega - \cos \omega)}{a^2(1 - 3 \cos^2 \omega + 2 \cos^3 \omega)}. \tag{1}$$

An analysis of this relation in the case of the small distortion shows that the largest difference in the value of θ_{hkl} occurs for the high-angle reflections with Miller indices

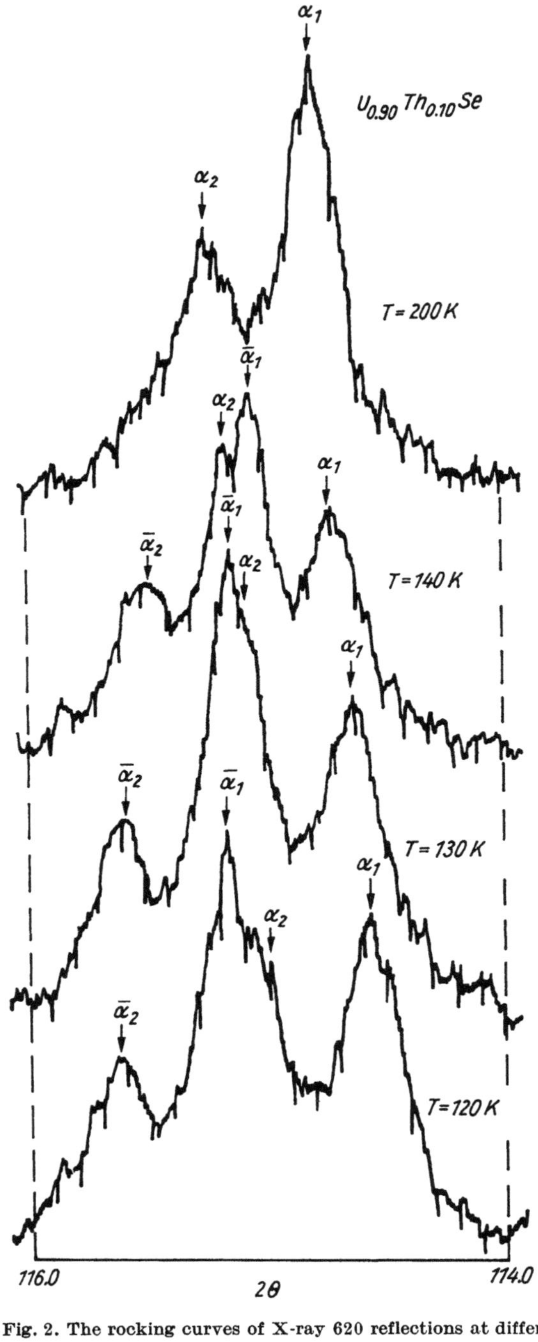

Fig. 2. The rocking curves of X-ray 620 reflections at different temperatures for $U_{0.90}Th_{0.10}Se$

$hk0$ and $\bar{h}k0$. A typical contour of the reflection 620 with and without a rhombohedral distortion of the cubic cell are presented in Figure 2 for sample $U_{0,90}Th_{0,10}Se$. In order to calculate the angle of distortion ε ($\varepsilon = 90 - \omega$) the Bragg reflections with Miller indices 420, 620 and sometimes 820 were analysed. As is seen in Figure 2 this rocking curve has two well resolved components α_1 and α_2 of Cu-radiation. Assignment of the measured position of the peaks to the $\bar{\alpha}_1$ and $\bar{\alpha}_2$ components for some reflections $hk0$ and $\bar{\alpha}_1$ and $\bar{\alpha}_2$ for $\bar{h}k0$ were checked by the relation:

$$\frac{\sin \theta_{hk0,\alpha_1}}{\sin \theta_{hk0,\alpha_2}} = \frac{\lambda_{\mathrm{CuK}\alpha_1}}{\lambda_{\mathrm{CuK}\alpha_2}} = 0.99752\,. \tag{2}$$

To calculate the angle ω one has to to compare the relations (1) for $hk0$ and $\bar{h}k0$.

For example for reflections 620 and $\bar{6}20$ this comparison gave:

$$\cos \omega = \frac{3u + 3 - \sqrt{196u^2 + 182u + 49}}{6u - 4} \tag{3}$$

where

$$u = \frac{\sin^2 \theta_{620,\alpha_1}}{\sin^2 \theta_{\bar{6}20,\alpha_1}}\,.$$

The difference between ω determined from the positions of the reflections 420 and 620 does not exceed 0.02°.

Sometimes for comparing the rhombohedral distortion with tetragonal one, caused by magnetic order along [100] axis, uses a quantity $(c' - a')/a'$, which is calculated from another type of the rhombohedral cell. The choice of new cell is illustrated in Figure 3. The interstitial angle, ω' is now close to 60°. In this cell c' is a distance along the trigonal axis and a' is a distance in the plane perpendicular to the trigonal axis such that for the cubic cell $c'/a' = 1$. According to LANDER and MUELLER the ratio $(c' - a')/a'$ may be calculated from the relation:

$$\frac{c' - a'}{a'} = -\frac{3}{2} \cdot \varepsilon \tag{4}$$

where $\varepsilon = 90 - \omega$ must be expressed in rad. If the [111] axis of the cube is shortened, $(c - a) < 0$ and $\varepsilon > 0$, and the reverse is true, if the [111] axis is stretched. The value of $(c' - a')/a'$ at various temperatures in the $U_xTh_{1-x}Se$ and $U_xY_{1-x}Se$ systems are presented in Figure 4a and Table 1, and in Figure 4b and Table 2, respectively.

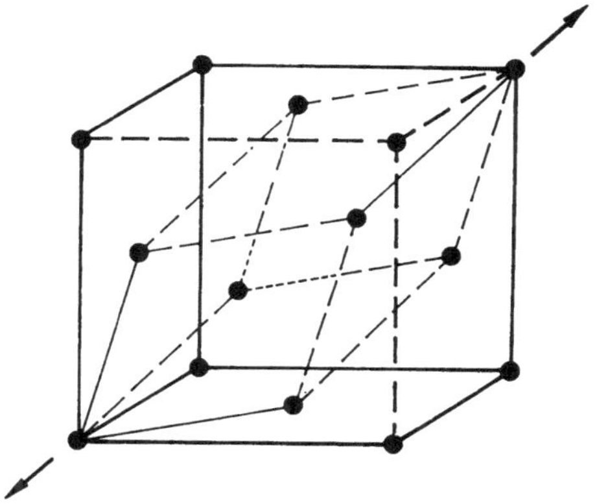

Fig. 3. Rhombohedral unit cell with $\omega' = 60°$ inside the face centred cubic cell with $\omega = 90°$

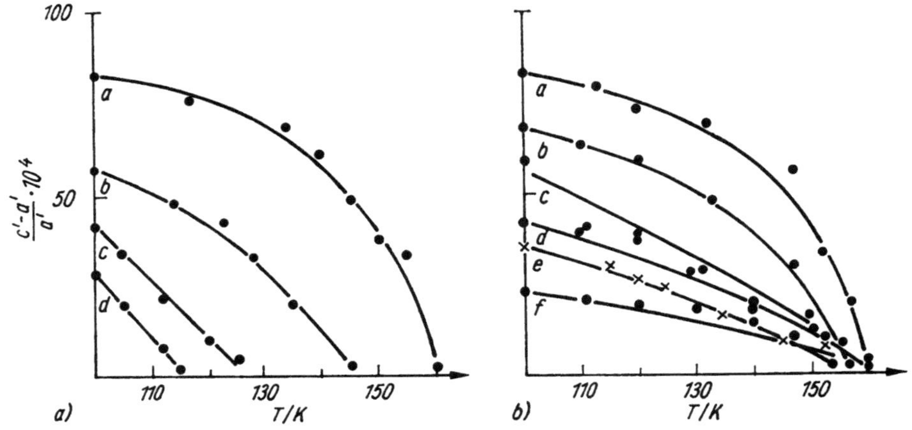

Fig. 4. a) Dependence of $\frac{c'-a'}{a'} \cdot 10^4$ on the temperature for $U_xTh_{1-x}Se$ samples; $x = 1.00$ (a), 0.90 (b), 0.85 (c), 0.80 (d); Dependence of $\frac{c'-a'}{a'} \cdot 10^4$ on the temperature for some $U_xY_{1-x}Se$ samples: $x = 1.00$ (a), 0.95 (b), 0.90 (c), 0.85 (d), 0.75 (e), 0.60 (f)

Table 1

Calculated values $\frac{c'-a'}{a'} \cdot 10^4(D)$ at various temperatures (T) for the $U_xTh_{1-x}Se$ system

$x = 1.00$		$x = 0.90$		$x = 0.85$		$x = 0.80$	
T/K	D	T/K	D	T/K	$D\Delta$	T/K	D
100	82	100	57	100	42	100	29
117	76	114	48	105	34	105	20
134	69	123	43	112	21	112	8
140	61	128	33	120	10	115	2
145	48	135	20	125	5		
150	38	140	3				
155	34						
160	3						

The accuracy of the calculated data is about $\pm 5 \cdot 10^{-4}$. So far no good theory of magnetostriction exists, and other authors used mainly two approaches. One, phenomenological, is based on assumptions elaborated by Akulov and Becker (MORRISH) and the other utilizes the crystal field theory together with the Stevens equivalent operators formalism (CALLEN, CALLEN). The former one gives a good agreement between theoretical and experimental results for d-metals and their compounds, the latter for lanthanides and their alloys. The situation in actinide compounds is more complex and the theory of their magnetic and related properties is based on Coqblin-Schriefer model (see FREEMAN and LANDER for a comprehensive review). Basing on the phenomenological approach of Akulov and Becker (MORRISH) it is possible to estimate the coefficients of magnetostriction λ_{111} and λ_{100} for cubic crystal using a simple, geometric construction. The relative change of [111] axis of the cubic cell may be written as:

$$\left(\frac{\Delta l}{l}\right)_{111} = \frac{3}{4}\lambda_{100}\left(\beta_1^2 + \beta_2^2 + \beta_3^2 - \frac{1}{3}\right) + \frac{3}{2}\lambda_{111}(\beta_1\beta_2 + \beta_2\beta_3 + \beta_1\beta_3) \qquad (5)$$

Table 2

Calculated values $\frac{c' - a'}{a'} \cdot 10^4 (D)$ at various temperatures (T) for $U_xY_{1-x}Se$ system

$x = 0.95$		$x = 0.90$		$x = 0.85$		$x = 0.80$	
T/K	D	T/K	D	T/K	D	T/K	D
100	68	100	49	100	42	100	40
110	62	120	40	110	40	108	38
120	59	128	39	120	37	115	37
133	48	140	28	129	29	124	30
147	30	150	18	140	19	130	27
153	10	160	1	151	13	140	19
158	1			155	8	150	12
						155	4
$x = 0.75$		$x = 0.70$		$x = 0.65$		$x = 0.60$	
T/K	D	T/K	D	T/K	D	T/K	D
100	36	100	30	100	25	100	23
115	30	110	22	112	21	111	21
120	26	121	19	119	19	120	20
125	24	130	17	125	17	130	18
135	16	140	16	135	15	142	15
146	14	150	7	145	9	147	11
152	9			150	2	150	3

where β_i are the direction cosines of the magnetisation. If the substance is strongly anisotropic and the easy axis of magnetisation is also [111] axis, then $\lambda_{111} \gg \lambda_{100}$ and the relation becomes much simpler:

$$\left(\frac{\Delta l}{l}\right)_{111} \approx 2\lambda_{111} \,. \qquad (6)$$

On the other hand it is possible to find changes in some direction in a rhombohedrally distorted cubic cell from simple trigonometry. In the crystal lattice with distorted NaCl-type structure the relative change along [111] axis may be written in the form:

$$\left(\frac{\Delta l}{l}\right)_{111} = \sqrt{3 - 4 \sin^2 \frac{\omega}{2}} - 1 \qquad (7)$$

where ω is an angle of the distorted cubic cell. This formula allows one to estimate the value of λ_{111}, which for USe and their solid solutions at 100 K was found to amount to $(1 \div 4) \cdot 10^{-3}$. The measured value of $\varepsilon = 90 - \omega$ calculated for USe at 100 K is $0.30 + 0.32°$ being in good agreement with values found by MARPLES, and MARPLES et al.).

4. Final conclusions

a) It was shown that X-ray diffraction measurements at low temperatures give a possibility to estimate a magnitude of magnetostriction using not only single crystals but also powder samples.

b) The distortion of the crystal cell was found to exist not only in USe but also in their solid solutions. A weakening of ferromagnetic exchange interactions caused by the diamagnetic dilution of USe lowers the magnitude of distortion of the crystal cell in a manner similar to the lowering of magnetisation (PALEWSKI; PALEWSKI et al.).
c) The estimated values of distortion $(\varepsilon, (c' - a')/a')$ are in good agreement with the data of other authors (MARPLES; LANDER, MUELLER).
d) The value of magnetostriction coefficient λ_{111} was calculated on the basis of the Akulov and Becker approach. The magnitude of λ_{111} does not exceed $(1 \div 4) \cdot 10^{-3}$ at 100 K.

References

ALDRED, A. T., de V. du PLESIS, P., LANDER, G. H.: J. Magn. Magn. Mat. **20**, 236 (1980)
ALLBUTT, M., DELL, R. M., JUNIKSON, A. R., MARPLES, J. A. C.: J. Inorg. Nucl. Chem. **32**, 2159 (1970)
CALLEN, E., CALLEN H.: Phys. Rev. **129**, 578 (1963)
FREEMAN, A. J., DARBY, J. B.: The Actinides: Electronic Structure and Related Properties, New York 1974
FREEMAN, A. J., LANDER, G. H.: Handbook on the Physics and Chemistry of the Actinides, New York, Amsterdam 1985
LANDER, G. H., MUELLER, M. H.: Phys. Rev. B **10**(5), 1994 (1974)
MARPLES, J. A. C.: J. Phys. Chem. Sol. **31**, 2431 (1970)
MARPLES, J. A. C., SAMPSON, C. F., WEDGWOOD, F. A., KUZNIETZ, M.: J. Phys. C, Sol. St. Phys. **8**, 708 (1975)
MORRISH, A. H.: Fizyczne podstawy magnetyzmu, Warszawa 1970
PALEWSKI, T.: phys. stat. sol. (a) **84**, K47 (1984)
PALEWSKI, T., WĘGLOWSKI, S., PAJAK, R.: Physica **86–88B**, 111 (1970)
PALEWSKI, T., WOJCIECHOWSKI, W.: to be published
TAKAHASHI, Y., WESTRUM, Jr., E. F.: J. Phys. Chem. **69**, 3918 (1965)
TRZEBIATOWSKI, W., SUSKI, W.: Bull. Acad. Pol. Sci., Ser. Sci. Chim. **10**, 399 (1962)
VOGT, O.: Physica **102B**, 206 (1980)
WEDGWOOD, F. A., KUZNIETZ, M.: Conf. Digest No 3, p. 168, Rare Earth and Actinides, Durham 1971

Received January 7, 1987)

Authors' addresses:

Dr. T. PALEWSKI
International Laboratory of High Magnetic Fields
and Low Temperatures
Ul. Prochnika 95
53-529 Wrocław, Poland

Dr. J. HORN
Institute of Inorganic Chemistry and Metallurgy of Rare Elements,
Technical University, Wybrzeże Wyspiauskiego 27
53-529 Wrocław, Poland

Cryst. Res. Technol.	22	1987	8	1055–1061

A. V. N. GUPTA, J. V. RAO, N. V. L. N. PRASAD, P. VENKATACHARYULU

Department of Physics, Nagarjuna University, Nagarjunanagar, India, and T.J.P.S. College, Guntur, Andhra Pradesh, India

Phase Transition Studies in the Liquid Crystal P-Cyanophenyl Trans-4-pentylcyclohexane Carboxylate

Investigations on specific volume and ultrasonic velocity measurements (at four different fixed frequencies) as a function of temperature are carried out in the liquid crystal p-cyanophenyl trans-4-pentylcyclohexane carboxylate. From these studies various physical parameters like adiabatic compressibility, molar sound velocity, molar compressibility, van der Waals constant and intermolecular free lengths are evaluated. The variation of the order parameter with temperature at the four different fixed frequencies is also studied and the results are discussed.

Introduction

Phase transition studies in liquid crystalline materials are very interesting and also of considerable importance in view of their established technological applications in recent years. The molecular properties like density, compressibility, viscosity, specific heat and also ultrasonic velocity as well as ultrasonic absorption exhibit anomalous behaviour at the phase transition temperatures. Ultrasonic velocity studies are widely adopted to investigate some of these properties and also these studies can provide additional information and confirmation about the different phases and phase transitions (NATALE; RAO et al. 1983; CHOUDARY et al.) as the liquid crystals exhibit anomalous behaviour in the vicinity of the phase transition.

A number of earlier investigators (GRAY; DUBIS, ZANN; BYRON et al.) investigated the properties of liquid crystals containing ester group. The present studies deal with the specific volume and ultrasonic behaviour of p-cyanophenyl trans-4-pentylcyclohexane carboxylate, which forms a part of our systematic investigations on the phase transitions of liquid crystals (RAO et al.).

Experimental

The liquid crystal

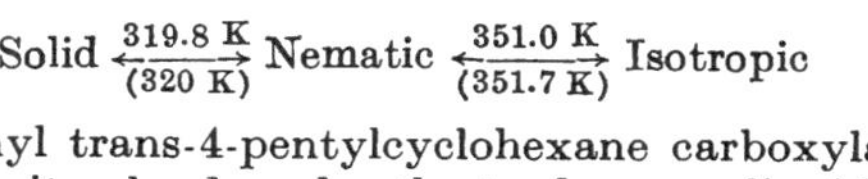

← (read-CN)

$$\text{Solid} \underset{(320\ \text{K})}{\overset{319.8\ \text{K}}{\longleftrightarrow}} \text{Nematic} \underset{(351.7\ \text{K})}{\overset{351.0\ \text{K}}{\longleftrightarrow}} \text{Isotropic}$$

p-cyanophenyl trans-4-pentylcyclohexane carboxylate was obtained from M/s. Hoffman La Roche, Switzerland, under the trade name liquid crystal component 1850. As the compound was obtained in its pure form and was subjected to experimental investigations without any further purification. The transition temperatures were measured using a polarizing microscope with a hot stage arrangement. The scheme of the transition temperatures are indicated above along with the company quoted values given in brackets.

The specific volume as a function of temperature in this compound is studied using a bi-capillary pycnometer with a bore diameter of 0.2 mm. The temperature dependence of the ultrasonic velocity is investigated at four different fixed frequencies i.e., 0.8 MHz, 2 MHz, 3 MHz, and 6 MHz using the multifrequency ultrasonic interferometer supplied by Mittal Enterprises, New Delhi, India. Further details regarding the experimental cell etc. are presented in detail elsewhere (VENKATACHARYULU et al.). The estimated accuracies in the measurement of specific volume, ultrasonic velocity, and temperature are ± 0.0001 cm³/g, $\pm 0.2\%$, and ± 0.1 K, respectively.

Results and discussion

The temperature dependence of the specific volume in the range 342—359 K for the liquid crystal 1850 is shown in Figure 1. From this figure it can be readily seen that the specific volume decreases linearly with the decrease of temperature both in the isotropic as well as in the nematic phases, except in the vicinity of isotropic-nematic transition. In the vicinity of N-I transition there is an abrupt increase in specific volume with increase of temperature. The increase in specific volume with temperature in the nematic phase can be well attributed to the decrease in the degree of order and thus decrease in the molecular packing with the increase of temperature.

The variation of ultrasonic velocity (V) with temperature in this mesogen is studied in the range 339—361 K at all the four different fixed frequencies and is represented in Figure 2. It can be readily seen from the figure that the ultrasonic velocity increases

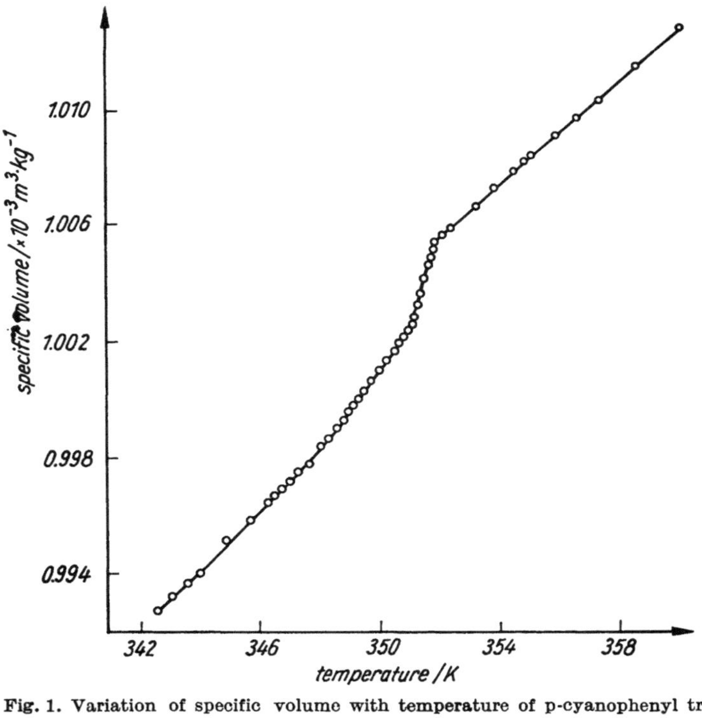

Fig. 1. Variation of specific volume with temperature of p-cyanophenyl trans-4-pentylcyclohexane carboxylate

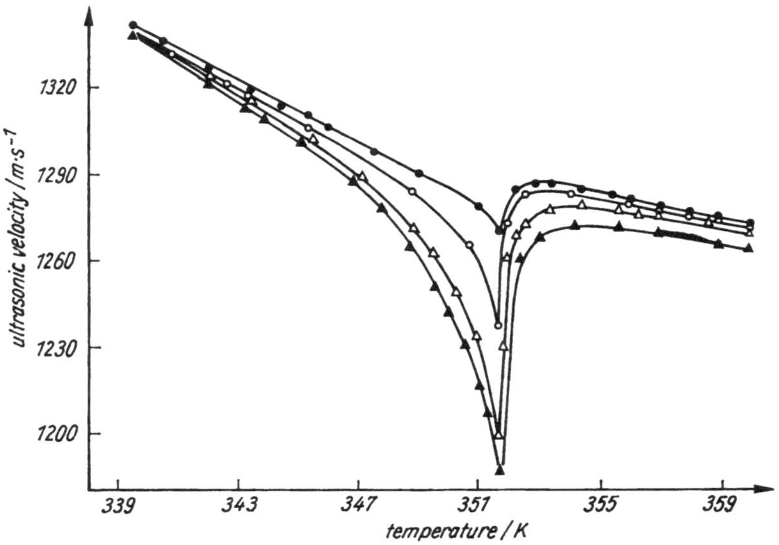

Fig. 2. Variation of ultrasonic velocity with temperature of p-cyanophenyl trans-4-pentylcyclohexane carboxylate. MHz: 6 (●), 3 (○), 2 (△), 0.8 (▲)

with decrease of temperature in the isotropic liquid phase and also in the nematic phase. There is an anomalous behaviour in the vicinity of isotropic-nematic transition exhibiting a rapid decrease in the ultrasonic velocity as the transition temperature is reached, attaining a minimum at the transition temperature and then again increasing rapidly after the transition. Using the specific volume (v) and the ultrasonic velocity (V) data, the physical parameters like adiabatic compressibility (β_{ad}), molar sound velocity (R_n), molar compressibility (B), van der Waals constant (b) and the mean free lengths (L) are calculated using the relations (Rao; Jacobson)

$$\beta_{ad} = v/V^2 \qquad R = Mv(V)^{1/3},$$

$$B = Mv/(\beta_{ad})^{1/7} \qquad b = Mv\left[1 - \frac{RT}{MV^2}\left\{\sqrt{1 + \frac{MV^2}{3RT}} - 1\right\}\right]$$

$$L = \left(\frac{k}{\beta_{ad}}\right)^{1/p}$$

where M is the molecular weight, k and p are constants which are, however, to some extent temperature dependent (Jacobson), S the order parameter (Maier, Saupe; Chandrasekhar, Madhusudhan) including its variation with temperature is estimated at the four different fixed frequencies.

The isotropic-nematic transition temperature of the compound from ultrasonic studies is found to be 351.7 K at all the four frequencies which is in close agreement with the value obtained from the specific volume studies. It can be readily observed from the Figure 2, in the isotropic phase the ultrasonic velocity increases with the decrease of temperature similar to ordinary liquids, as the temperature is decreased the mean distance between the molecules decreases, leading to an increase in the potential energy of interaction and thus to the increase in velocity. Just above the transition temperature the ultrasonic velocity decreases with the decrease of temperature attaining a minimum at the isotropic-nematic transition and a further decrease in temperature results again to an abrupt increase in the ultrasonic velocity and then increases linearly with the decrease of temperature. The entire course of this anomalous behaviour takes place within 4 K. As the ultrasonic velocity is related to the molecular structure, intensity and character of molecular interaction (BAYER, LETCHER; CANDAU, LETCHER) the sudden decrease in the velocity in the vicinity of the isotropic-nematic transition can be attributed to the sudden structural changes that arise from the disordered isotropic liquid phase to the ordered nematic phase. The pretransitional effects occur on both sides of the isotropic-nematic transition temperature at all the four frequencies which can be explained on the basis of FRENKEL's heterophase fluctuation theory. But the pretransitional effects are predominant on the lower side of the transition i.e., nematic side and in literature such cases are plenty (RAO et al. 1983; BAHADUR). The pretransitional effects on the isotropic side can be attributed to the small heat of transition from the nematic phase to the isotropic liquid.

Figure 3 represents the variation of the adiabatic compressibility of this compound with temperature at the four different fixed frequencies which also indicates the pretransitional effects on both sides of the transition. However, they are more predominant on the lower side, i.e. the nematic side of the transition. Just as in the case of ultrasonic velocity the pretransitional effects decrease with the increase of frequency. The anomalous effects are also more at lower frequencies when compared to higher frequencies. The value of the adiabatic compressibility is less in the nematic phase when compared to the isotropic phase. But in the specific volume measurements the pretransitional effects are mainly observed on the lower side of the transition as was reported in a number of cases by the earlier investigators (RAO et al.; BAHADUR).

The molar sound velocity or Rao number (R_n) and the molar compressibility (Tab. 1) show interesting results. It can be observed that the RAO number is almost constant in each state with a high value in the isotropic state when compared to the nematic state. From the same figure it can also be observed that at a particular temperature the Rao number increases with increasing frequency both in the isotropic and nematic

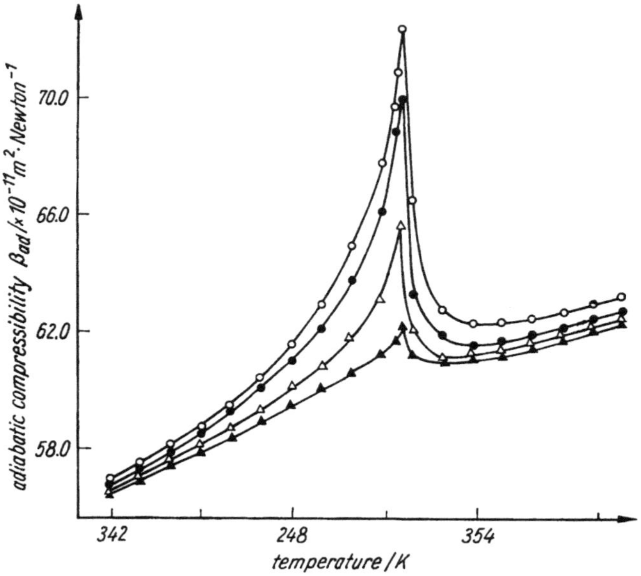

Fig. 3. Variation of adiabatic compressibility with temperature of p-cyanophenyl-trans-4-pentyl-cyclohexane carboxylate. MHz: 0.8 (○), 2(●), 3 (△), 6 (▲)

Table 1
Molar sound velocity (R_n) and molar compressibility (B) at different temperatures of p-cyanophenyl trans-4-pentylcyclohexane carboxylate

S. No.	temperature (K)	0.8 MHz		2 MHz		3 MHz		6 MHz	
		R_n	B	R_n	B	R_n	B	R_n	B
1.	359.0	3276	8078	3281	8687	3283	8691	3284	8693
2.	358.0	3276	8675	3280	8685	3282	8690	3282	8691
3.	357.0	3274	8673	3279	8682	3280	8686	3282	8690
4.	356.0	3271	8668	3277	8678	3279	8683	3281	8687
5.	355.0	3270	8663	3275	8675	3279	8683	3280	8685
6.	354.0	3267	8655	3273	8670	3277	8678	3278	8682
7.	353.0	3259	8638	3267	8656	3274	8671	3276	8677
8.	352.0	3225	8561	3251	8620	3261	8643	3268	8659
9.	351.7	3177	8450	3195	8491	3230	8567	3260	8633
10.	351.5	3194	8489	3200	8504	3233	8579	3259	8638
11.	351.0	3203	8510	3215	8538	3240	8595	3257	8634
12.	350.0	3220	8549	3229	8571	3246	8610	3258	8635
13.	349.0	3232	8577	3238	8590	3250	8619	3258	8635
14.	348.0	3240	8595	3243	8602	3253	8624	3258	8635
15.	347.0	3246	8607	3247	8611	3254	8628	3258	8635
16.	346.0	3249	8616	3251	8620	3256	8631	3258	8637
17.	345.0	3252	8621	3254	8626	3258	8635	3259	8639
18.	344.0	3254	8627	3256	8631	3259	8639	3260	8641
19.	343.0	3256	8631	3259	8637	3260	8641	3261	8643
20.	342.0	3258	8636	3251	8641	3261	8643	3262	8645

R_n expressed in units of [cm^3 $(m/sec)^{1/3}$]; B expressed in units of [cm^3 $(dyne/cm^2)^{1/7}$]

phases. However, the behaviour of Rao number with temperature is similar to the ultrasonic velocity. Just as in the case of ultrasonic velocity the pretransitional effects are predominant on the lower side, i.e. nematic side of the transition and found to be more at lower frequencies. The behaviour of the molar compressibility is exactly similar to the Rao number as reported by the earlier workers (RAO et al. 1983; BAHADUR; RAO et al.) over a number of thermotropic liquid crystals.

The van der Waals constant (b) is found to be the same at all the four different frequencies both in the isotropic phase as well as in the nematic phase and its value is slightly less in the nematic phase when compared to its value in the isotropic phase. The value of b in the isotropic phase at a temperature of 358 K at the four frequencies is 290.60, 290.20, 290.70, and 290.00 and the value of the same in the nematic phase at a temperature 343 K is 286.08, 286.11, 286.10, and 286.12 at the same four frequencies. This indicates that the actual volume of the molecules in a gram mole is constant for that particular phase as the van der Waals constant represents four times the volume of the molecules in a gram mole.

The intermolecular free lengths (L) are calculated in the temperature range 359 K—342 K at all the four different fixed frequencies in this compound adopting the JACOBSON's theory and are represented in Table 2. The intermolecular free lengths decrease with decrease of temperature both in the isotropic and nematic states. Just like the ultrasonic velocity, adiabatic compressibility etc. the intermolecular free lengths also exhibit anomalous behaviour in the vicinity of the transition temperature. The value of L is found to be less at higher frequencies when compared to its value at lower frequency at any particular temperature.

The order parameter at the transition temperature is estimated using the specific volume data adopting MAIER-SAUPE tables and the variation of the order parameter with temperature is estimated adopting the method suggested by CHANDRASEKHAR

Table 2
Intermolecular free lengths ($L \times 10^{11}$ m)

S. No.	temperature	0.8 MHz	2 MHz	3 MHz	6 MHz
1	359.0	5.928	5.909	5.900	5.897
2	356.0	5.872	5.852	5.841	5.833
3	354.0	5.741	5.711	5.696	5.688
4	353.0	5.677	5.643	5.614	5.602
5	351.7 (T.C.)	6.040	5.958	5.801	5.669
6	350.0	5.786	5.743	5.666	5.616
7	348.0	5.638	5.623	5.582	5.559
8	346.0	5.480	5.472	5.451	5.444
9	343.0	5.466	5.458	5.450	5.440
10	342.0	5.374	5.364	5.360	5.366

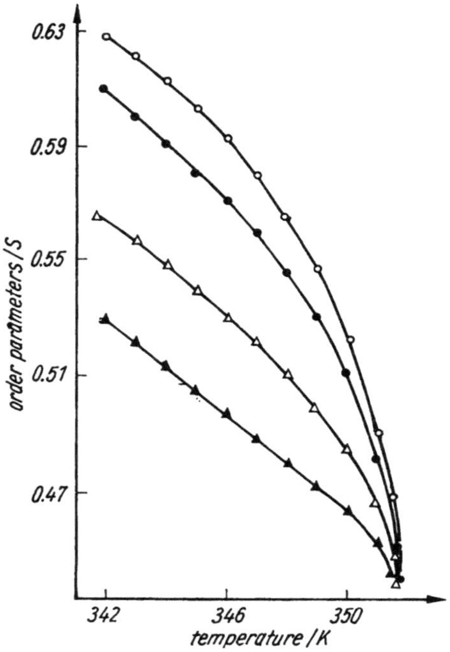

Fig. 4. Variation of order parameter with temperature of p-cyanophenyl trans-4-pentylcyclohexane carboxylate. MHz: 0.8 (○), 2 (●), 3 (△), 6 (T▲)

and Madhusudhana at the four different frequencies and is presented in Figure 4. It can be observed from the figure that the order parameter increases with the decrease of temperature indicating an increasing order with the decrease of temperature and decreases with the increasing frequency. The decrease in the order parameter with the increase of frequency may be due to the fact as the frequency increases the disturbance in the molecules also increase which leads to a decrease in the order parameter.

References

Bahadur, B.: Acoustica **34**, 486 (1975)
Bahadur, B.: Acoustica **33**, 277 (1975)
Bahadur, B.: J. Phys. (c), Solid State Physics **9**, 11 (1976)

BAYER, R. T., LETCHER, S. V.: Physical ultrasonics, New York 1969
BYRON, D. J., LACEY, D., WILSON, R. W.: Mol. Cryst. Liq. Cryst. **45**, 267 (1978)
CANDAU, S., LETCHER, S. V.: Advances in Liquid Crystals **3**, 168 (1978)
CHANDRASEKHAR, S., MADHUSUDHANA, N. V.: Liq. Cryst. — Part II, edited by G. H. BROWN, New York 1969, p. 291
CHOUDARY, L. V., RAO, J. V., MURTY, P. N., MURTY, C. R. K.: phys. stat. sol. **74**, 431 (1982)
DUBIS, J. C., ZENN, A.: J. Phys. **37**, C3, No. 3, 35 (1976)
FRENKEL, J.: Kinetic Theory of Liquids, New York 1946
GRAY, G. W.: Mol. Cryst. Liq. Cryst. **31**, 275 (1975)
JACOBSON, B.: Acta Chemica Scavandicavica **6**, 1485 (1952)
JACOBSON, B.: J. Chem. Phys. **20**, 927 (1952)
MAIER, W., SAUPE, A.: Z. Naturforsch. **159**, 287 (1960)
NATALE, G. G.: J. Acoust. Soc. Amer. **65**, 1265 (1978)
RAO, J. V., RAO, K. R. K., CHOUDARY, L. V., VENKATACHARYULU, BALIAH, V.: Phase Transitions (in press)
RAO, K. R. K., RAO, J. V., VENKATACHARYULU, BALIAH, V.: Mol. Cryst. Liq. Cryst. (in press)
RAO, K. R. K., RAO, J. V., CHOUDARY, L. V., VENKATACHARYULU, P., BALIAH, V.: Z. Phys. Chem. (in press)
RAO, K. R. K., RAO, J. V., CHOUDARY, L. V., MURTY, C. R. K.: Acta Physics Polonia A **63**, 419 (1983)
RAO, M. R.: Ind. J. Phys. **14**, 109 (1940)
RAO, M. R.: J. Chem. Phys. **9**, 682 (1941)
VENKATACHARYULU, P., GUPTA, A. V. N., RAO, J. V.: Cryst. Res. Technol. **22**, **743** (1987)

(Received January 7, 1987)

Author's address:

c/o Dr. P. VENKATACHARYULU
Department of Physics, Nagarjuna University
Nagarjunanagar-522510, India

Cryst. Res. Technol.	22	1987	8	1063–1072

M. T. SEBASTIAN, P. KRISHNA

Cochin University, Cochin, India,
and Banaras Hindu University, Varanasi, India

X-ray Investigation of the Mechanisms of Phase Transformations in Single Crystals of ZnS, $Zn_xCd_{1-x}S$, and $Zn_xMn_{1-x}S$ (II)

Comparison of Observed and Calculated Diffraction Effects

The experimentally observed intensity profiles recorded from ZnS, $Zn_xCd_{1-x}S$ and $Zn_xMn_{1-x}S$ crystals at different stages of the 2H–3C and 2H–6H transformations are compared with those computed theoretically from the three parameter model in the previous paper. It is found that the 2H–3C and 2H–6H transformations occur by the non-random nucleation of deformation faults. The probability of the growth of a nucleus is much greater than that of the creation of a fresh nucleus. At least two different fault probabilities have therefore to be employed in computing the diffraction effects in each case. The transformation behaviour of $Zn_xCd_{1-x}S$ and $Zn_xMn_{1-x}S$ crystals is strongly influenced by variations in stoichiometry (x). Large values of x ($x > 0.95$) favour the 2H–3C transformation, whereas smaller values ($x \leqq 0.94$) favour the formation of the 6H phase. A comparison of the calculated and observed intensity distributions indicates that in some crystals the 2H–3C and 2H–6H transformations occur simultaneously in different regions of the same single crystal.

Introduction

In the previous paper (SEBASTIAN, KRISHNA 1987) we developed a theoretical three-parameter model to describe the mechanism of the 2H–3C and 2H–6H transformations observed in ZnS and its solid solutions with MnS and CdS. The model is based on the observation that there is a marked enhancement of X-ray diffuse intensity in 6H positions in crystals which undergo the 2H–3C transformation and conversely there is a marked enhancement of X-ray diffuse intensity in 3C positions in crystals undergoing the 2H–6H transformation. However, two possibilities exist for explaining this observation and a decision between them can only be made by comparison with experiment:

(i) The 2H–3C and 2H–6H transformations may be taking place simultaneously in different regions of the crystal in which case the resultant diffracted intensity would be a weighted sum of the intensities due to the two transformations calculated separately (by the respective two parameter models) or

(ii) The transformations may occur homogeneously throughout the volume of the crystal with one set of fault probabilities α, β, and γ characterizing the 2H–6H transformation and a different set of fault probabilities α, β and γ characterizing the 2H–3C transformation. In this case the observed intensity would correspond to that calculated from the three-parameter model for both transformations.

In order to distinguish between these two possibilities we have performed experiments by quenching crystals undergoing the 2H–3C and 2H–6H transformations at different stages of the transformation and then recorded the diffuse X-ray intensity produced by the intermediate disordered state along the 10.*L* reciprocal lattice row. For this the quenched crystals were mounted on a four-circle computer-controlled single crystal diffractometer and the diffuse intensity along the 10.*L* row recorded. This paper compares the diffraction effects so recorded with those theoretically predicted for the two transformations in different situations.

Experimental Results

Single crystals of ZnS, $Zn_xCd_{1-x}S$ and $Zn_xMn_{1-x}S$ used in the present study were grown from the vapour phase in the presence of H_2S gas (SEBASTIAN, KRISHNA 1983). An X-ray diffraction study showed (SEBASTIAN et al. 1982) that the ZnS crystals contain 2H, 3C or polytype structures and their mixtures. Many crystals contained a high concentration of stacking faults on their basal plane, as revealed by the presence of continuous diffuse streaks along reciprocal rows parallel to c^*. The concentration of stacking faults varies not only from crystal to crystal but also along the length of the same crystal. Needle-shaped crystals of ZnS, $Zn_xCd_{1-x}S$, and $Zn_xMn_{1-x}S$, 1 – 2 mm length and 0.1 to 0.4 mm diameter, were annealed for 1 hour each in vacuum at successively higher temperatures in the range 300 to 1100 °C and then quenched into cold water. The crystals when annealed are found to give rise to progressive coloration due to diffusion of impurities from the walls of the silica tube on which they are resting and vaporize at high temperatures. So we followed the method of ADDAMIANO and AVEN of keeping the crystals in immersed pure ZnS powder inside an evacuated silica tube. By the use of this method the crystals remained clear even after annealing at 1100 °C for 2 hours. After each annealing run the silica tube was immersed in cold running water to arrest the structural transformations taking place. The crystals did not show the appearance of any kinks or shear on examination under a vickers projection microscope. The crystals showed numerous parallel strips or striations perpendicular to the *c*-axis. The structure of the crystals was re-examined by X-ray diffraction after each annealing run. It is found that the transformation commences with the insertion of basal plane stacking faults as indicated by the diffuse streak developed along the 10.*L* row. Figure 1 depicts the 10.*L* reciprocal lattice row of a disordered 2H $Zn_{0.9}Mn_{0.1}S$ crystal as recorded on a 15° *C*-axis oscillation photograph after successive annealing runs in the

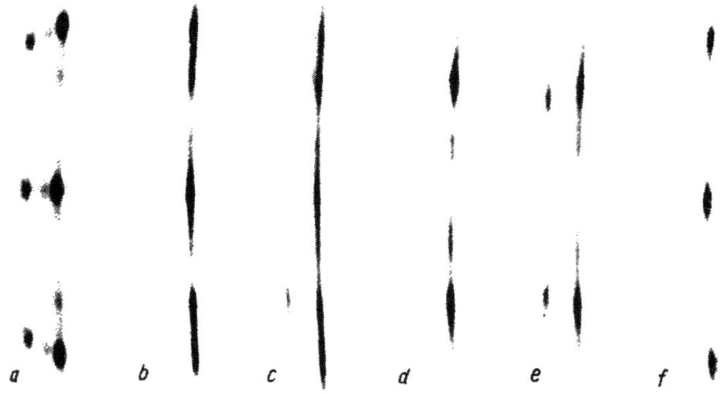

Fig. 1. The 10.*L* reciprocal lattice row of a disordered 2H $Zn_{0.9}Mn_{0.1}S$ crystal as recorded on 15° *C*-axis oscillation photographs after annealing at successively higher temperatures for 1 hour each and quenching at (a) room temperature (b) 300 °C, (c) 400 °C, (d) 750 °C, (e) 1000 °C, (f) 1080 °C (camera radius 3 cm, CuK_α radiation)

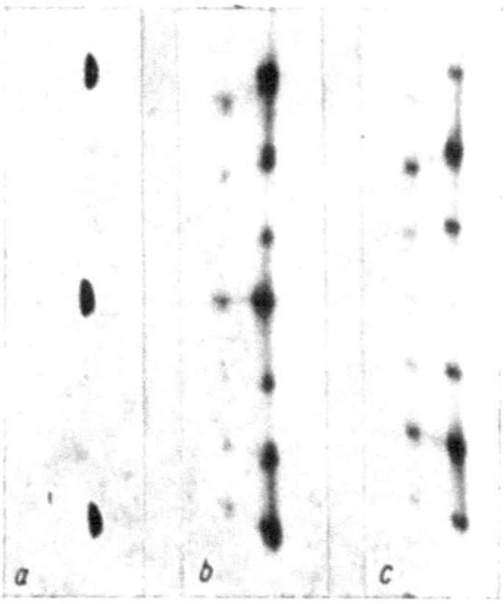

Fig. 2. The 10.L reciprocal lattice row of a 2H $Zn_{0.94}Cd_{0.06}S$ crystal recorded on 15° at C-axis oscillation photographs after annealing at successively higher temperatures for 1 hour and quenching; (a) room temperature, (b) 800 °C, (c) 900 °C (camera radius 3 cm, CuK_α radiation)

range 300 to 1080 °C for 1 hour each. This crystal shows a 2H–3C transformation with intensity enhancement near the 6H positions in the intermediate states. Figure 2 shows the 10.L reciprocal lattice row of a 2H $Zn_{0.94}Cd_{0.06}S$ crystal recorded before and after annealing and quenching at 800 °C. This crystal shows a 2H–6H transformation with inten-

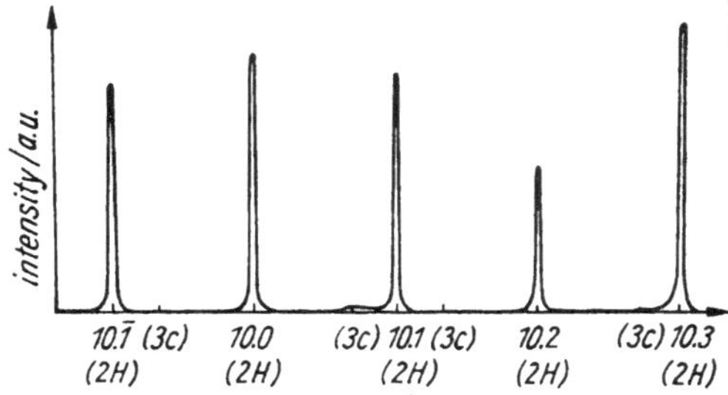

3)

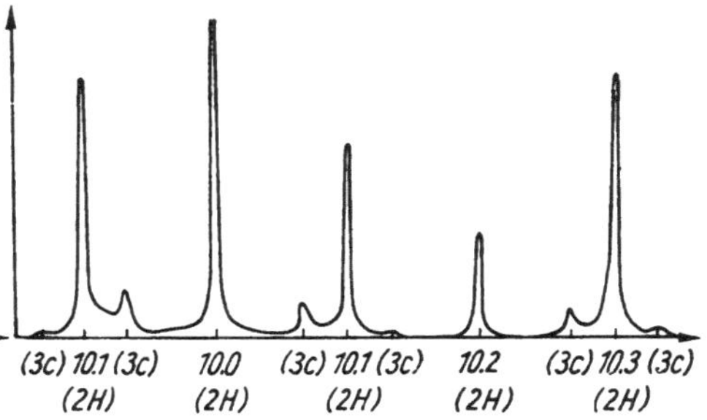

4)

Fig. 3. Single crystal diffractometer plot of intensity versus L along the 10.L reciprocal lattice row of an as-grown 2H ZnS crystal

Fig. 4. Single crystal diffractometer plot of intensity versus L along the 10.L reciprocal lattice row recorded after annealing a perfect 2H ZnS crystal at 400 °C for 1 hour

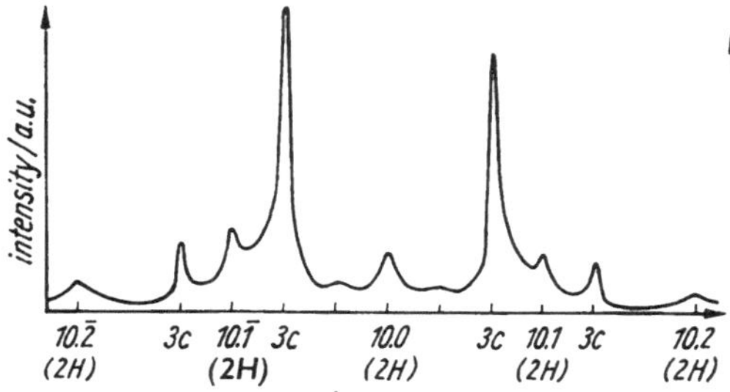

5)

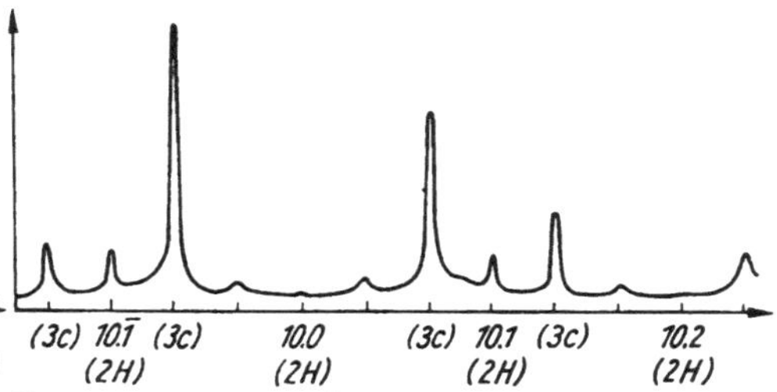

6)

Fig. 5. Single crystal diffractometer record of intensity versus L along the 10.L row of a ZnS crystal nearly transformed to a disordered 3C structure obtained by annealing a 2H crystal at 600 °C for 1 hour. The vertical lines indicate the 6H positions

Fig. 6. Single crystal diffractometer record of intensity versus L along the 10.L row of a ZnS crystal transformed to a disordered 3C structure obtained by annealing a 2H crystal at 700 °C for 1 hour. The vertical lines indicate the 6H position

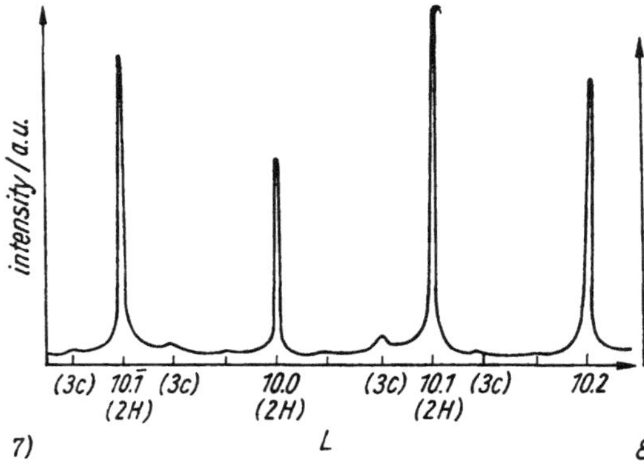

Fig. 7. Single crystal diffractometer record of intensity versus L along the $10.L$ row of a disordered 2H $Zn_{0.98}Mn_{0.02}S$ crystal recorded after annealing a perfect 2H crystal at 700 °C for 1 hour. The vertical lines indicate 6H positions

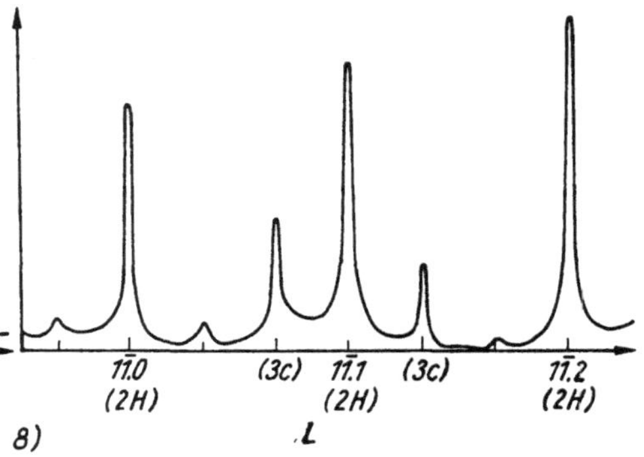

Fig. 8. Single crystal diffractometer record of intensity versus L along the $1.\bar{1}L$ row of an as-grown 2H $Zn_{0.9}Mn_{0.1}S$ crystal partially transformed to the 6H structure. The vertical lines indicate 6H positions

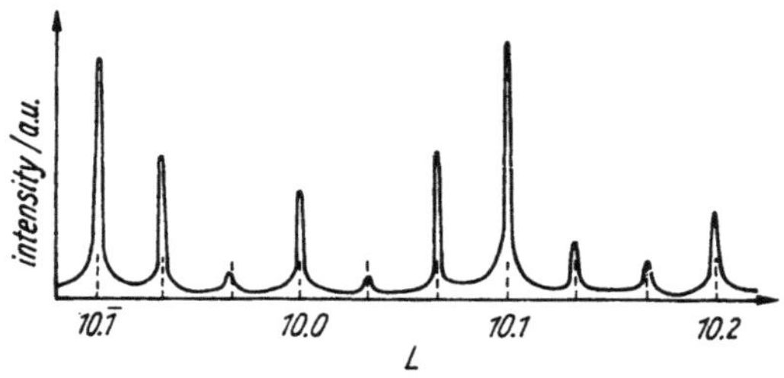

Fig. 9. Single crystal diffractometer record of intensity versus L along the $10.L$ row of a 2H $Zn_{0.94}Cd_{0.06}S$ crystal recorded after annealing a perfect 2H crystal at 800 °C for 1 hour. The vertical lines indicate the 6H positions

sity enhancement near the 3C positions. The point intensity distribution along the $10.L$ (or the symmetry related $1\bar{1}.L$) row of several partially transformed ZnS, $Zn_xCd_{1-x}S$ and $Zn_xMn_{1-x}S$ crystals quenched at different stages of the 2H–3C and 2H–6H transformations was recorded in steps of $\Delta L = 0.01$ using MoK_α radiation and a computer controlled four circle single crystal diffractometer. The crystal and the counter were held stationary during each observation. The divergence of the incident beam was adjusted to cover the mosaic spread in the crystal. The sharp reflections with $H-K = 0$ mod 3 were used to orient the crystal. Figure 3 to 6 show the observed intensity distribution along the $10.L$ row of ZnS crystals at four different stages of the 2H–3C transformation. Figure 7 and 8 show the observed intensity distribution along the $10.L$ and $1\bar{1}.L$ row of two $Zn_xMn_{1-x}S$ crystals recorded in an initial and an advanced stage of the 2H–6H transformation, respectively. Figure 9 shows the intensity distribution along the $10.L$ row of a $Zn_xCd_{1-x}S$ crystal in an advanced stage of the 2H–6H transformation.

These records suffer from a gradual change in the scale-factor with the result that the relative intensities at L values far removed from each other cannot be compared with accuracy but the individual peak profiles, peak positions and half-widths can be ascertained with reasonable accuracy for a meaningful comparison with theoretically predicted values. This problem will now be discussed in some detail.

Problems in recording directionally diffuse intensity

It was observed by PANDEY and KRISHNA that the diffuse $10.L$ reflections produce widely different intensity records on the diffractometer on rotating the crystal through different angles (ψ) around the scattering vector. This is because the surface of the Ewald Sphere always has a finite thickness on account of the divergence in the incident beam and the length of the $10.L$ row which is intercepted by the Ewald Sphere changes with ψ, as shown in Figure 10.

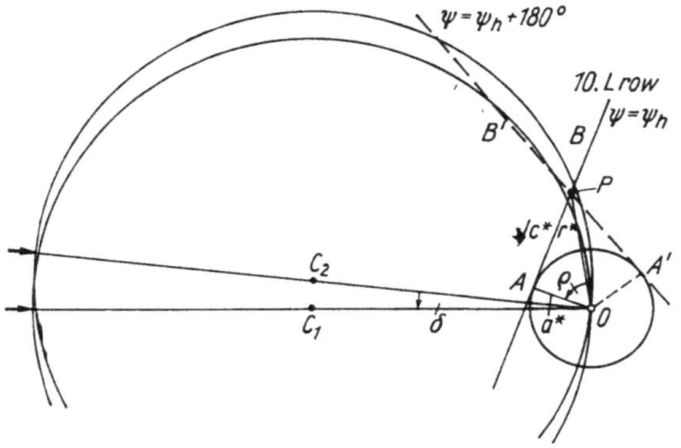

Fig. 10. Ewald's construction showing that different lengths 'ΔL' of the streak are diffracted at different ψ values. OA and OA' represent the directions of the a^* vector for $\psi = \psi_h$ and $\psi = \psi_h + 180°$. The diffuse $10.L$ reciprocal lattice row is perpendicular to a^*. The point being recorded is P. Due to the divergence δ in the incident beam, very different lengths ΔL of the row will be in diffracting condition in the two cases

Recently MÜLLER has shown that one can make the $10.L$ row to lie in the horizontal diffracting plane by using a value of $\psi = \psi_h$ given by

$$\tan \psi_h = \frac{c_x^* \sin \chi_B \sin \varphi_B - c_y^* \sin \chi_B \cos \varphi_B + c_z^* \cos \chi_B}{c_x^* \cos \varphi_B + c_y^* \sin \varphi_B}$$

for the Enraf-Nonius geometry. The symbols have their usual meanings. It should be noted that φ and χ are in reverse directions in the Philips and Enraf Nonius diffractometers. The angles χ_B and φ_B are fixed by the diffractometer for ensuring the bisecting geometry $\omega = \theta$ which defines the position $\psi = 0$. c_x^*, c_y^* and c_z^* are the components of c^* along the diffractometer axes (x along the incoming beam, z vertical and upwards and y perpendicular to both these). c_x^*, c_y^* and c_z^* can be calculated from the direction cosines in the orientation matrix. Therefore one can always calculate the value of ψ_h and work with the $10.L$ row in the diffraction plane as shown below. Figure 11 shows the reflection circle in the horizontal diffracting plane with a choice of reciprocal lattice vectors $a \cdot a^* = b \cdot b^* = c \cdot c^* = \lambda$. The ω-scan rotates the crystal about a vertical axis through 0. If the crystal is rotated by an angle $\Delta\omega$ (scan range) then the triangle OAB shifts to the dotted position OAB. Now the $10.L$ row cuts the reflection circle at a different L value which means a different θ value. When ω changes, the angles θ and ϱ both will change. It is easy to see from the figure that

$$\Delta\omega = \Delta\varrho + \Delta\theta\,. \tag{1}$$

Dividing by ΔL we get

$$\frac{\Delta\omega}{\Delta L} = \frac{\Delta\varrho}{\Delta L} + \frac{\Delta\theta}{\Delta L}. \tag{2}$$

From triangle OAB

$$L^2 c^{*2} = a^{*2} + (2 \sin \theta)^2\,.$$

Differentiating with respect to L and rearranging we get,

$$\frac{\Delta\theta}{\Delta L} = \frac{L c^{*2}}{2 \sin 2\theta}. \tag{3}$$

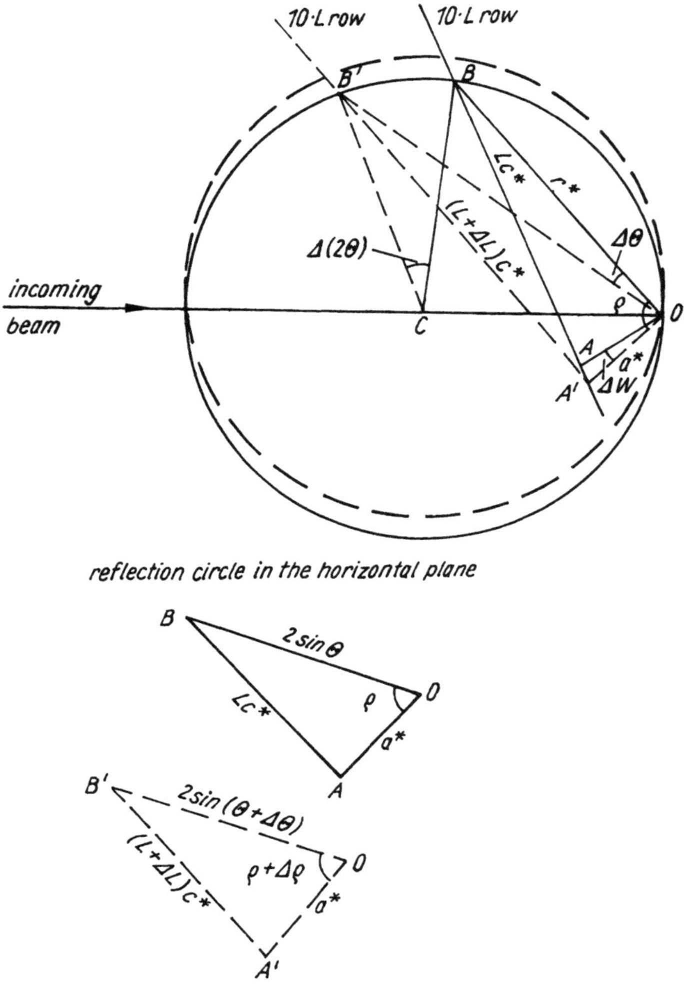

Fig. 11. Ewald's construction showing the intersection of the diffuse 10.L reciprocal lattice row with the Ewald sphere when the row lies in the horizontal plane and the crystal is rotated through an angle $\Delta\omega$. If the crystal is held stationary and the incident beam has a divergence $\delta = \Delta\omega$, the Ewald sphere would move to the dotted position producing a similar effect

But $\tan \varrho = \dfrac{\mathrm{AB}}{\mathrm{OA}} = \dfrac{\mathrm{L}c^*}{a^*}$

$$\frac{\Delta\varrho}{\Delta L} = \frac{c^*}{a^*}\cos^2\varrho\,. \tag{4}$$

Substituting equations (3) and (4) in (2) we get,

$$\frac{\Delta\omega}{\Delta L} = \frac{c^*}{a^*}\cos^2\varrho + \frac{Lc^{*2}}{2\sin 2\theta} \tag{5}$$

where

$$\cos \varrho = \frac{a^*}{2 \sin \theta}.$$

From equation (5) we can calculate the ΔL corresponding to the scan range ($\Delta\omega$). If the stationary crystal-stationary counter method is used to record the point intensity then $\Delta\omega$ is equivalent to the angular divergence δ of the incident beam as seen from the crystal. Thus knowing the length of the streak ΔL intercepted by the Ewald sphere, it is possible to calculate the correct intensity by dividing by ΔL, provided ΔL is not large. For this it is necessary to ensure that one is not working close to the geometry where the $10.L$ row makes tangential contact with the Ewald sphere. When $\psi = \psi_h + 90°$ the triangle OAB will be vertical and the divergence will cause a very large ΔL to get diffracted with intensity extending also in the vertical directions perpendicular to the detector slit. To minimize these effects, in the present investigation we have selected such [reciprocal lattice rows (with $H-K \neq 0 \bmod 3$) for which ψ_h does not vary much in going from one reflection to another along the row and is small (less than 10°). The intensity along the entire row can then be recorded with the row very nearly horizontal. The length of the streak entering the detector slit was monitored by taking polaroid photographs in front of the detector at different points along the row. A gradual variation in ΔL with L will not affect the peak positions and the half widths of individual peaks and this data can be used in the calculations.

Comparison of the observed diffraction effects with those computed theoretically

A comparison of the experimentally recorded intensity profiles in different stages of the 2H–3C and 2H–6H transformations with those computed theoretically using the three parameter model with different values of α, β and γ shows that the model is not applicable as such to these crystals. During the 2H–3C transformation the theory predict the shift of the intensity of the those 6H reflections which do not coincide with the cubic peaks ($L = \pm 1/3, \pm 4/3$) towards the nearby 3C positions. The experimentally recorded profiles do not show any shift of these reflections (see Fig. 5 to 9). The peak broadening for these crystals is also not in good agreement with the

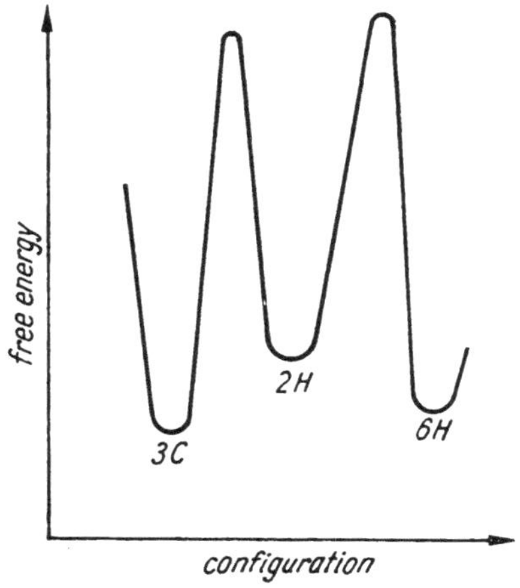

Fig. 12. Schematic free energy diagram for simultaneous transformation from a metastable 2H structure to 3C and 6H structure

predicted values. We therefore conclude that the 2H–3C and 2H–6H transformations do not occur simultaneously in the same region of the crystal. Such a possibility could exist if the free energy diagram at the temperature of anneal was of the kind shown in Figure 12. In this case the metastable phase 2H could transform simultaneously to the 3C and 6H phases in different proportions in conformity with the three parameter model.

Since this is not the case, it appears that a few regions of the 2H crystal transform to a 3C structure while the rest of the crystal transforms to 6H. In this case the resultant intensity distribution would be a weighted superposition of the intensity distributions calculated separately for the 2H–3C and 2H–6H transformations using the two parameter models.

Discussion of results

The 2H-3C and 2H-6H transformations in ZnS, $Zn_xCd_{1-x}S$ and $Zn_xMn_{1-x}S$ occur through the non-random nucleation of deformation faults. The probability distribution for both transformations are best described by the two parameter models which assume that the probability of the growth of nuclei is much greater than the probability of formation of a fresh nucleus. The observed enhancement of intensity in 6H positions during the 2H–3C transformation and in 3C positions during the 2H–6H transformation is due to the two transformations occurring simultaneously in different regions of the same crystal. This may be attributed to the local variations in the stoichiometry (or impurity content) in the crystals. It is known that the addition of small quantities of CdS or MnS stabilizes the 6H phase, consequently a different proportion of transformation to the 3C and 6H structure in different regions of the same crystal is quite possible in the solid solution phases though it is a bit surprising in ZnS itself. SINGER also observed the presence of such independently diffracting domains in ZnS crystals; the independent scattering from such domains being additive. The enhancement of diffracted intensity near the 6H positions was also observed by FLEET in mineral sphalerite (3C) ZnS crystals which were believed to have formed from the wurtzitep hase. AKIZUKIO bserved 2H, disorder, 3C, 4H and 15R, structures in different regions of heated ZnS minerals by the lattice imaging technique using an electron microscope.

All the crystals (over 150) on which we have made the annealing studies did not show any macroscopic kinks or shear on the external shape of the crystals. The transformation proceded by a disordering process as revealed by the intensification of diffuse streaks along the $10.L$ reciprocal lattice rows. The twin reflections of the cubic structure are of comparable intensity. All these features are different from those expected by the periodic slip (screw dislocation) mechanism. In fact we borrowed some crystals from Prof. Steinberger and did the annealing studies. It is found that the crystals showed kinks on partially transformed crystals (see Fig. 13). The kinked region is untwinned and perfect as revealed by the sharpness of the cubic ($10.\frac{2}{3}$ and $10.\bar{4}/3$) reflections (see Fig. 14). The cubic reflections are misoriented with respect to the faint diffuse streak, due to the fact that the diffuse row is from the straight 2H regions and the cubic reflections from the kinked region. These characteristics is clear picture of crystals undergoing transformation by the periodic slip mechanism (STEINBERGER).

A number of investigators (FULLER, BRIDGMAN; FREY; KULLERUD; VAN WIERINGIN) have studied the effect of crushing on the 2H–3C transformation and reported that the hexagonal modification inverts to cubic when crushed. SMITH and HILL (1956)

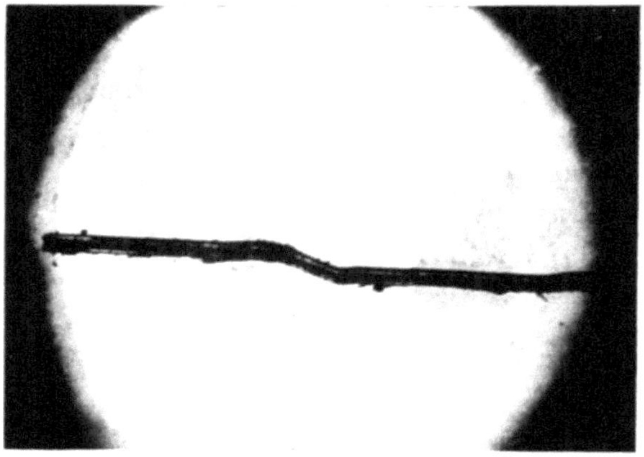

Fig. 13. The optical micrograph of a crystal showing a kink

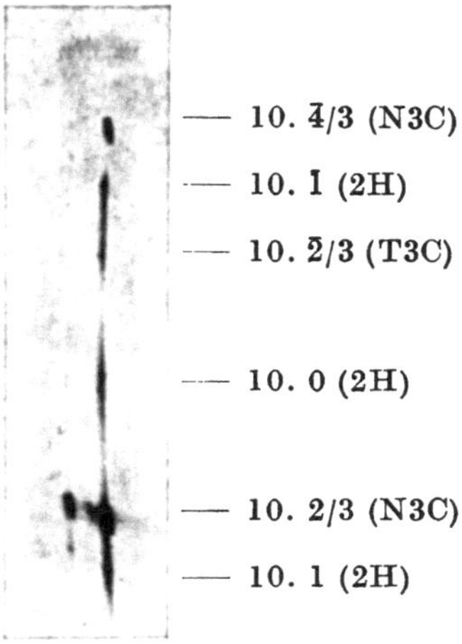

Fig. 14. The 10.*L* reciprocal lattice row recorded from the kinked region of the crystal shown in Figure 13

however, observed that crushing merely randomizes the structure since from both 3C and 2H starting materials a similar product is obtained. SHACHAR et al. reported that single crystals 2H ZnS transform to a disordered 3C + 6H on applying mechanical stress at room temperature by a needle. FARKHAS-JAHNKE and GACS have shown that single crystals of 4H ZnS transform to a disordered twinned 3C structure on pressing the crystals in a line perpendicular to the *c*-axis. Recently DUROSE et al. reported that ball milling of CdS and CdSe randomises the structure by deformation faulting. An examination of their diffractograms shows the broadening of the sharp reflections like 11.0, 11.2, 00.2 etc. which can be due to strains induced by the grinding process. It is difficult to see how a random and violent process as ball milling or grinding could introduce precise and specific dislocation movements. Undoubtedly dislocations are produced, but equally certainly their movements in general are not likely to be ordered. The grinding or ball milling methods cannot be applied to study phase transformation since it will randomise the structure, powdering of the sample and introduce considerable strains in the crystal.

A considerable amount of work has been done in the past using powder diffraction methods to study the nature of faults and the fault concentration by observing line widths and peak shifts. The information so obtained gives values that are averaged over all the grains in a polycrystalline sample. Such a method is obviously unsuitable for the study of polytypic materials where different grains could possess different crystal structure. Therefore only single crystal studies can provide reliable information about the disorder in individual polytypic crystals. The assumption of random disorder breaks down when such disorder occurs in the course of a phase transformation. A suitable model of non-random faulting based on the mechanism of the transformation has therefore to be employed to interpret the experimental results. This affords an indirect means of obtaining information about the mechanism of the transformation.

It should, however, be noted that the theoretical calculations described in the previous paper are based on a model which applies only approximately. By lattice imaging a partially transformed ZnS crystal in the electron microscope AKIZUKI found evidence of stacking faults that did not extend all the way across the crystal. The spacing between "layers" in ZnS is not exactly the same at a fault. The spacing of cubic close-packed layers is known to differ from that of hexagonal close-packed layers by 0.3%. Similarly the contribution of strains within the crystal may not always be negligible since ZnS crystals get easily deformed (SHACHAR et al.). All these factors, if present in large measure, can cause deviations between the observed and calculated diffraction profiles.

The authors are grateful to the Department of Science and Technology, New Dehli, for financial assistance and to prof. Steinberger for giving us a few ZnS crystals containing screw dislocations.

References

ADDAMIANO, A., AVEN, M.: J. Appl. Physics **31**, 36 (1960)
AKIZUKI, M.: Amer. Mineralogist **66**, 1006 (1981)
BRIDGMAN, P. W.: Amer. J. Sci. **237**, 7 (1939)
DUROSE, K., FELLOWS, A. T., BRINKMAN, A. W., RUSSEL, G. J., WOODS, J.: J. Mater. Sci. **20**, 3783 (1985)
FARKHAS-JAHNKE, M., GÂCS, P.: Kristall und Technik **14**, 1475 (1979)
FLEET, M. E.: Amer. Mineralogist **62**, 540 (1977)
FREY, F.: Ann. Physik. **2**, 147 (1948)
FULLER, M. K.: Phil. Mag. **8**, 658 (1929)
KULLERUD, G.: Norsk. Geol. Tidesskr. **32**, 61 (1953)
MÜLLER, V. U.: Acta Cryst. **A35**, 957 (1979)
PANDEY, D., KRISHNA, P.: J. Phys. **D10**, 2057 (1977)
SEBASTIAN, M. T., PANDEY, D., KRISHNA, P.: phys. stat. sol. **71A**, 633 (1982)
SEBASTIAN, M. T., KRISHNA, P.: Solid State Commun. **49**, 879 (1983)
SEBASTIAN, M. T., KRISHNA, P.: J. Cryst. Growth **66**, 586 (1984)
SEBASTIAN, M. T., KRISHNA, P.: Cryst. Res. Technol. **22**, 929 (1987).
SHACHAR, G., MARDIX, S., STEINBERGER, I. T.: J. Appl. Physics **39**, 2485 (1968)
SINGER, J.: Act. Cryst. **16**, 601 (1963)
SMITH, F. G., HILL, V. G., Acta Cryst. **9**, 821 (1956).
STEINBERGER, I. T., in: Crystal Growth and Characterizatian of polytype structures, Bd. P. KRISHNA, New York 1983.
VAN WIERINGIN, J. S.: Physica **19**, 397 (1953)

(Received October 11, 1986)

Authors' addresses:

Dr. M. T. SEBASTIAN
Lab. de Mineralogie-cristallographie
Université Pierre at Marie Curie
4 Jussieu
15252 PARIS, cedex 05, France

Professor P. KRISHNA
Banaras Hindu University, physics department
Varanasi 221 005, India

Cryst. Res. Technol.	22	1987	8	1073–1077

C. C. DESAI, M. S. V. RAMANA

Department of Physics, Sardar Patel University Vallabh Vidyanagar, Gujarat, India

Dielectric Properties of Ferroelectric Lead Hydrogen Phosphate Single Crystals

The ferroelectric lead hydrogen phosphate (LHP) single crystals have been grown by employing the gel technique. The controlled reaction between lead nitrate and orthophosphoric acid solutions by diffusion process in gel medium has been used. The dielectric constant (ε') and the dielectric loss (ε'') have been measured for the first time as a function of frequency in the range 10^2 to 10^4 Hz. They have also been determined as a function of temperatures in the range 300 to 680 K. The polarisation effect contributing to the dielectric constant has been studied. It has been observed that the mechanism of the dielectric behaviour is different in the lower and higher temperatur and frequency ranges. The activation energy of the oscillators has been calculated and found to be 1.12 eV. Attempt has been made to draw some qualitative conclusions, taking in view, the existing theories of various kinds of polarization and implications have been discussed.

1. Introduction

A detailed study of the dielectric properties of crystalline solids gives a good insight into the electric field distribution within them. From the measurements of the dielectric constant and the dielectric loss as a function of frequencies and temperatures, the various polarisation mechanism in solids can be understood and the defect concentration estimated approximately. Considerable amount of work along these lines has been carried out on a variety of materials, for example, alkali halides (LEHFELDT; WAGNER, HANTEMANN; KELTING, WITT; GOVINDA, RAO), on tin iodides (DESAI, RAI), on mercury iodide (REDDY, SIRDESHMUKH) and on transition metal oxides (RAO, SMAKULA), yielding valuable information. Such investigations also help in detecting structural transitions in crystalline solids because abrupt changes in dielectric properties take place at these transitions (SURYTH, HITCHCOCK).

Lead Hydrogen Phosphate (LHP) crystals are of considerable interest, particular for their basic studies of their ferroelectric, piezoelectric, optical and other pertient characteristics. They are often used in transducers and in many linear and non-linear devices. These crystals are also useful for testing microscopic theory of ferroelectricity. As far as is known to the authors, the dielectric properties of LHP have not been reported in the literature. It is the aim of the present paper to describe the results of our detailed and systematic investigations on the dielectric properties of LHP single crystals grown in our laboratory from a solution technique.

2. Experimental

2.1. Crystal growth

LHP single crystals were grown using the controlled reaction between lead nitrate and orthophosphoric acid by a diffusion process in silica gels. The production of LHP crystals is in accordance with the following chemical reaction:

$$Pb(NO_3)_2 + H_3PO_4 \rightarrow PbHPO_4 + 2\,HNO_3\,. \quad (1)$$

Experimental system and procedure for preparation of gels are described by the authors (DESAI, RAMANA) elsewhere. The experimental optimum conditions giving the best crop of transparent crystals with regard to their size and perfection were found to be:

gel density	= 1.04 gm cm^{-3}
gel *p*H	= 3.5
gelling time	= 25–30 hrs
concentration of $Pb(NO_3)_2$	= 2.0 M
concentration of H_3PO_4	= 1.0 M
growth temperature	= 22–26 °C
gel volume	= 200 ml.

The large crystals grew in about 2–3 weeks. The crystals were harvested with the physical removal of gels. The crystals were critical examined by chemical analysis, SEM, EDAX, X-ray diffraction, density measurement, thermogravimetric analysis and were confirmed to be LHP. They belong to monoclinic system and have lattice parameters: $a = 5.7822$ Å, $b = 6.6454$ Å, $c = 4.683$ Å, $\beta = 97.14°$ and its space group is P2/c and have 2 molecules per unit cell.

2.2. Dielectric measurements

LHP single crystals were mounted in the [001] direction between the two flat silver electrodes and silver paste was applied on both the experimental surfaces for firm contact. Pure silver (99.9%) wires were attached to these faces when the past was still wet and soft. It may be mentioned that the measurements were repeated on samples obtained from the same growth run. The experiments were conducted on well faceted small crystals of LHP. The geometrical dimensions of the samples were measured using a travelling microscope of least count 10^{-4} cm.

The dielectric measurements and dielectric loss have been carried out using a "GR 1620" type capacitance bridge in the frequency range 10^2 to 10^4 Hz. The capacitance readings were recorded when the values became steady. The measured values of capacitance were changed into permittivity ε using the relation:

$$\text{Permittivity } \varepsilon = \frac{Cd}{A} \quad (2)$$

where

C represents the capacitance of the crystal,
d is the thickness of the crystal, and
A is the area of opposite faces in contact with electrodes.

It was then enclosed in a resistance heated furnace and the temperature of the sample was increased slowly. The actual temperature of the sample was monitored using a chromel alumel thermocouple (±1 °C). The accuracy of measurements in dielectric constant ε' is 3% and that in the dielectric loss ε'' is about 6%.

3. Results and discussion

Figures 1a and 1b represent the variation of the dielectric constant ε' and loss ε'' at room temperature as a function of frequency respectively. The dielectric constant has a high value at low frequencies and it decreases with the increased in frequency and at about 10^4 Hz it becomes frequency independent. The dielectric loss decreases with frequency and it is of the order of 10 in the region where dielectric constant is independent of frequency.

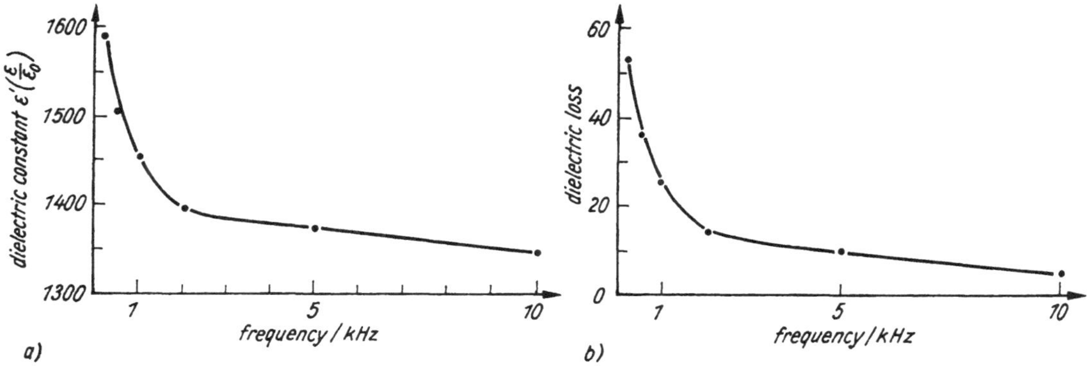

Fig. 1. a) Variation of the dielectric constant ε' of LHP single crystals with frequency at temperature 30 °C; b) Variation of ε'' of LHP crystals, with frequency at temperature 30 °C

The dielectric constant of a material is, usually, composed of four contributions: electronic, ionic, orientational and space charge polarisations. All these may be operative at low frequencies. In fact, the plot of dielectric constant with frequency (Fig. 1a) indicates which contributions are present. Thus, the increase of dielectric constant at lower frequencies is due to space charge polarization, which depends on the purity and perfection of the crystals. This conclusion seems to be borne out by the moderately large values for dielectric loss at low frequencies (SURYANARAYANA et al.).

The results of the variation of dielectric constant at different temperature at a fixed frequency of 10^4 Hz are summarized in Table 1. The plots of dielectric constants, and the dielectric loss ε'' at various temperatures are shown in Figures 2a and 2b respectively. The temperature has a complicated influence on the ε' and ε'' increasing the temperature is likely to decrease the number of oscillations per unit volume. The increase of atomic distances influences both electronic and ionic polarization.

The plots of ε' and ε'' initially show a slower increase upto a temperature of 523 K which may be attributed to the atomic polarizability and the expansion of lattice. The subsequent exponential increase of dielectric constants above 523 K

Table 1

Dielectric parameters of LHP crystals at different temperature at a fixed frequency 10^4 Hz

Obs. No.	temperature (T) of the crystal (K)	permittivity ε of the crystals ($\times 10^{-12}$ F cm^{-1})	dielectric constant $\varepsilon' = \varepsilon/\varepsilon_0$
1	313	119.02	1344.31
2	378	123.88	1399.22
3	423	128.09	1446.74
4	473	132.30	1494.26
5	523	136.51	1541.78
6	568	143.05	1615.71
7	623	156.61	1768.83
8	673	219.12	2481.64

$\varepsilon_0 = 8.854 \times 10^{-12}$ F m^{-1}

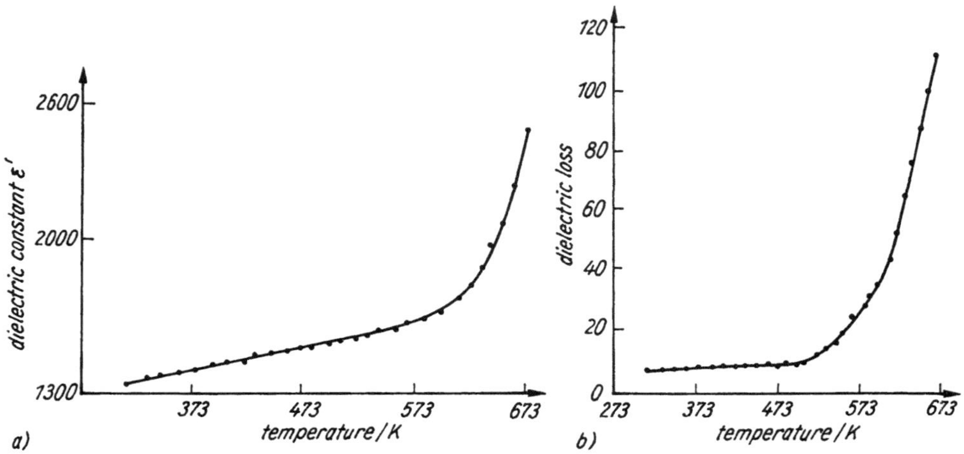

Fig. 2. a) A plot of dielectric constant ε' against various temperatures at 10^4 Hz frequency; b) Variation of ε'' with temperatures at a fixed frequency of 10^4 Hz

may be ascribed to space charge polarizations of thermally generated charge carriers (LAL, SRIVASTAVA).

The relaxation time (t) for the space charge polarization of thermally generated charge carriers in the studied temperature range, 300 to 680 K is given by the equation (GEVERS, DUPRE):

$$t = t_0 \exp (H/kT) \tag{3}$$

where H is the activation energy of dielectric oscillators.

From the graph of $\ln \varepsilon'$ against $1/T$ (Fig. 3) the activation energy of these oscillators in crystals has been evaluated and was found to be 1.12 eV.

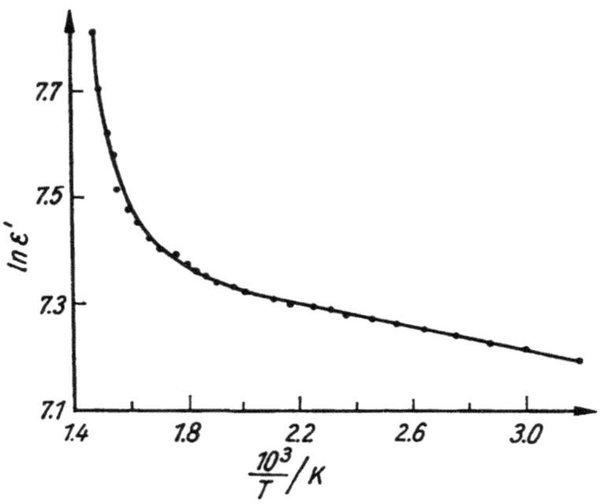

Fig. 3. A plot of $\ln \varepsilon'$ against $1/T$ at 10^4 Hz frequency

4. Conclusions

1. LHP single crystals upto $5 \times 3 \times 2\ mm^3$ in size have been grown using the gel techniques. They belong to the monoclinic system and having space group $P2/c$.

2. The dielectric constants and dielectric loss of LHP crystals have been determined at a frequency range of 10^2 to 10^4 Hz. They have also been determined in the temperature range of 300 to 680 K.
3. The different polarizations that contributed to the dielectric constants of LHP crystals have been identified from the nature of the variation of the dielectric constants with frequencies and temperatures.

The authors are thankful to Council of Scientific and Industrial Research (CSIR), New Delhi, for the sanction of the project, entitled "Growth and characterization of Ferroelectric crystals" (Project Number: 3(564)84 EMR-II, September 1984). The authors are grateful to Dr. G. K. Shivakumar for helpful discussion. The authors thank Mr. A. C. Desai and Mrs Lorra (London) for their keen interest in the present work.

References

DESAI, C. C., RAI, J. L.: J. Mater. Sci. Lett. **17**, 3249 (1982)
DESAI, C. C., RAMANA, M. S. V.: XVII National Seminar on Crystallography, University of Madras **22** (1985)
GEVERS, M., DUPRE, F. K.: Trans. Faraday Soc. **42A**, 47 (1946)
GOVINDA, S., RAO, K. V.: Phys. Status Solidi **27**, 639 (1975)
KELTING, H., WITT, H.: Z. Phys. **126**, 697 (1949)
LAL, H. B., SRIVASTAVA, K. G.: Can. J. Physics **47**, 3 (1969)
LEHFELDT, W.: Z. Phys. **85**, 717 (1933)
RAO, K. V., SMAKULA, A.: J. Appl. Physics **36**, 3953 (1965)
REDDY, Y. R., SIRDESHMUKH, L.: Solid State Comm. **51**, 407 (1984)
SURYANARAYANA, P., ACHARYA, H. N., RAO, K. V.: J. Mater. Sci. Lett. **3**, 21 (1984)
SURYTH, C. P., HITCHCOCK, C. S.: J. Amer. Soc. **54**, 4631 (1932)
WAGNER, G., HANTEMANN, P.: J. Chem. Phys. **18**, 72 (1950)

(Received January 2, 1987)

Author's address:

Professor C. C. DESAI, Mr. M. S. V. RAMANA
Department of Physics, Sardar Patel University
Vallabh Vidyanagar 388120
Gujarat State, India

Cryst. Res. Technol.	22	1987	8	1079–1084

C. C. DESAI

Department of Physics, Sardar Patel University, Vallabh Vidyanagar,
Gujarat, India

A New Technique for Measuring Thickness of Thin Solid Films

A new technique has been developed for measuring the thickness parameter for any thin solid films. Various thin films of silver, aluminium and antimony have been prepared by using vacuum evaporation plant at the vacuum of 10^{-5} torr. The films have been scratched by a fine metal pin. By employing a semi-silvered cylindrical convex lens, a point contact has been established between a silvered scratch-film and a cylindrical lens. Multiple beam Fizeau interference fringes have been obtained over the surface of thin films. Across the step of the scratch, because of height difference, the shift in the fringes have been observed. Due to the axis of point contact, no ambiguity in allocation of ringe order has been obtained. The thicknesses of silver, aluminium and antimony thin films have been determined very precisely. The technique gives high resolution with a good degree of reproducibility and implications are discussed.

1. Introduction

Thickness of thin films is very important parameter, as it is related to their physical properties and now a days such films are extensively used in science and in industry. Number of methods have been mentioned by TOLANSKY (1955, 1960) and HEAVENS by which the thickness of thin films may be measured but they have their own limitations. They have reported that the multiple beam interferometric method of measuring film thickness, is the most accurate method and this has been demonstrated by securing extremely fine multiple beam Fizeau fringes across a step formed on a substrate by a thin silver film.

For measuring the thickness of a thin film deposited on a substrate by interferometry, usually a monitor plate which is half covered with a thin mica sheet is kept side by side the substrate when the thin film is deposited on it. Thus the thickness of the film deposited on the monitor plate and the substrate is assumed to be the same. The mica sheet is removed from the monitor plate after a deposition of silver and the step so formed due to the film is measured after depositing a thin layer of silver on the monitor plate and then matching it against a silvered flat to produce multiple beam Fizeau fringes. Across the steps, because of height difference, the fringes are displaced. From the displacement of fringes, the height of the step is measured which gives the required thickness. This method has the following drawbacks:

(i) The thickness of the film is not directly measured, but the thickness of the film on the monitor plate is measured. (ii) In the multiple beam interferogram, it is very difficult to assign order of interference to the fringes and hence to judge which fringe has been displaced across the step.

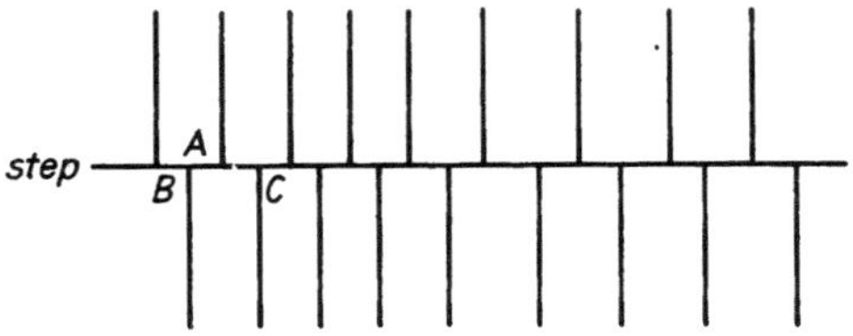

Fig. 1. Fizeau fringes across a step

TOLANSKY (1955) has shown that because the silver diffuses in the mica, there is a continuity of the fringes across the step and hence no ambiguity arise regarding the displacement of fringes. However, when the step is very sharp, ambiguity does arise. As for example, consider displacement of fringes across the step as shown in Figure 1. It is indeed very difficult to say whether A has been displaced to B or C or to any other fringe position on the lower half of the picture on which the thickness of the film depends. No doubt, if one locates the same order fringe on both sides of the step, the thickness can be measured precisely by this method, but allocation of order is not easy in this case. Efforts were made, therefore, to develop a suitable technique for measuring thickness parameter and the present paper describes the technique that was evolved. The uncertainty in allocation of orders mentioned above and its limitations can be removed.

2. Experimental procedure and results

In this method, there is no necessity to deposit the film on the monitor plate along with the substrate as described above. The thickness of the thin film deposited on a substrate can be measured directly as follows:

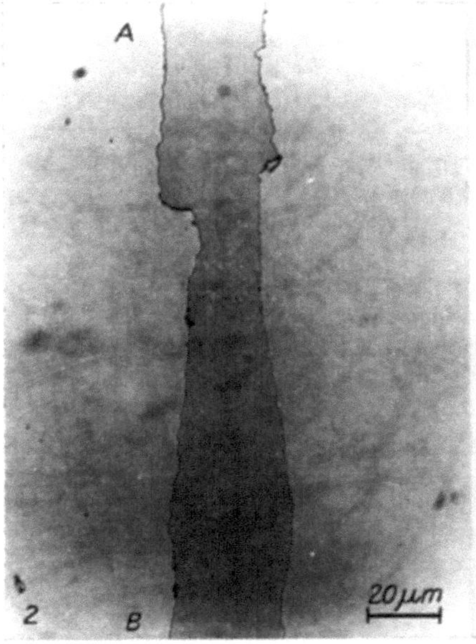

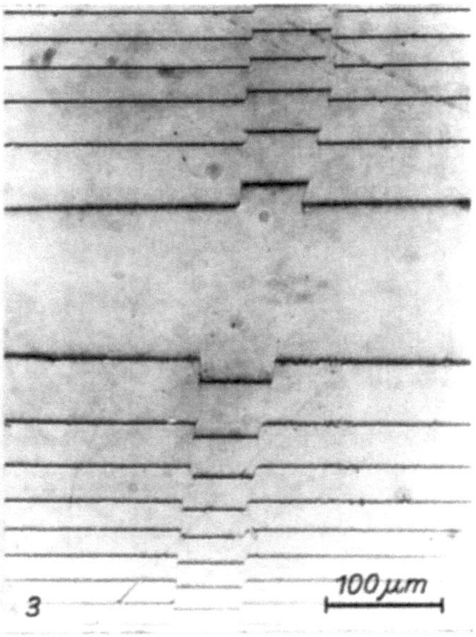

Fig. 2. A thin film of silver deposited on the glass substrate. AB marked is a scratch

Fig. 3. Multiple beam interferogram of thin film shown in Figure 2

The film deposited on the substrate the thickness of which is to be measured, is to be scratched with a sharp point, exactly in the region where the thickness is required to be known accurately. For drawing a scratch, the material should be so selected that it does not scratch the substrate but only cuts the film, e.g. if thin film of silver is deposited on glass, fine metal pin can scratch the film without scratching the glass. This is shown in Figure 2. The scratched marked by AB is then silvered by depositing about 200 Å thick film of silver in a vacuum coating unit, at the vacuum of 10^{-5} torr. Then it is matched against a semi-silvered cylindrical glass lens with its radius of curvature accurately known. The cylindrical surface of the lens (3 inch radius) should be so placed that its axis is at right angles to the scratch (AB) drawn.

The multiple beam Fizeau fringes in this set up will be parallel to the point of contact, i.e. parallel to the axis of the cylindrical lens with a shift at the site of the scratch. This is shown in Figure 3 which is a multiple beam interferogram taken on Figure 2, revealing the topography of the face of thin film. It is taken with 5461 Å line of mercury. Attention may be drawn to the following:

(i) The fringes have no structures between two lines indicating that the films are quite flat.
(ii) The scratch marked AB makes a sharp step of the thickness of the thin film.
(iii) The point of contact is clearly seen.
(iv) The fringes in the scratch will shift toward the lower order.
(v) There is no ambiguity in allocation of fringe orders, if the film thickness is less than $\lambda/2$ and hence no ambiguity in the measurement because the first fringe is the first order fringe.
(vi) By measuring the shift in the fringes the film thickness can be determined. The thickness parameter can be measured at the exact position where one desires to evaluate.

3. Calculations of thickness of thin films

In order to calculate the thickness of the film, set-up should be so arranged that the fringes run perpendicular to the scratch. The point of contact can be obtained by measuring the distance between the two first order fringes on the two sides and dividing it by two. Having known the point of contact, the perpendicular distance (D_n) of a number of fringes from it can be obtained. Since the location perpendicular to the axis of a cylinderical lens is a circle, the Newton's ring formula can be employed to measure the film thickness from the measured displacement of fringes within the scratch as follows:

For any ring of fringe order (n), we have

$$n\lambda = 2t \tag{1}$$

where λ represents the wavelength, 5461 Å line of mercury, t is height (thickness) of the scratch and n the fringe order. Any irregularity arises from a local change in t, say dt, and this produces a change in fringe order dn such that

$$\text{Film thickness} \quad t = dn\left(\frac{\lambda}{2}\right). \tag{2}$$

Thus to evaluate change in microtopography, it is only necessary to measure dn. These sharp fringes have a width some one-fiftieth of an order, and it is possible to detect displacements of one-fifth of a fringe width, hence the limit value of dn which is measurable is one two hundred and fiftieth of an order. The fringe distances were measured using a filar micrometer eye piece fitted to a metallurgical Vickers projection microscope at a constant magnification of 200. An accuracy of 1 micron can be achieved with this method. The data of perpendicular distance (D_n) of a number of fringes from the axis of contact is summarized in Table 1.

Table 1
Measurements of various fringe orders

Fringe order number (n)	distance of the fringe order from the axis of contact (D_n) in mm	square of the distance $(D)^2$ in $mm^2 \times 10^{-2}$
1	0.081	0.65
2	0.112	1.25
3	0.134	2.40
4	0.155	2.80
5	0.174	3.02
6	0.194	3.75
7	0.204	4.20
8	0.221	4.90
9	0.235	5.50
10	0.248	6.18
11	0.260	6.79
12	0.268	7.20
13	0.282	8.00
14	0.295	8.70

A plot of D_n^2 (square of the distance of fringe orders from the axis of contact) against n (fringe order number) is shown in Figure 4. To measure the shift of the first order within the scratch, measure the distance of first order fringe within the scratch as D_1' and subtract it from D_1 the distance of the displaced first order fringe from the axis of contact outside the scratch. Let this be X. Obtain values of X^2 and from the graph (Fig. 4) determine n corresponding to X^2, which will give the fraction of the order displacement. Now that D_1 corresponds to the thickness ($\lambda/2$) the thickness corresponding to the fraction of an order displacement can be calculated. As an example, the measurements made on the fringes are described below:

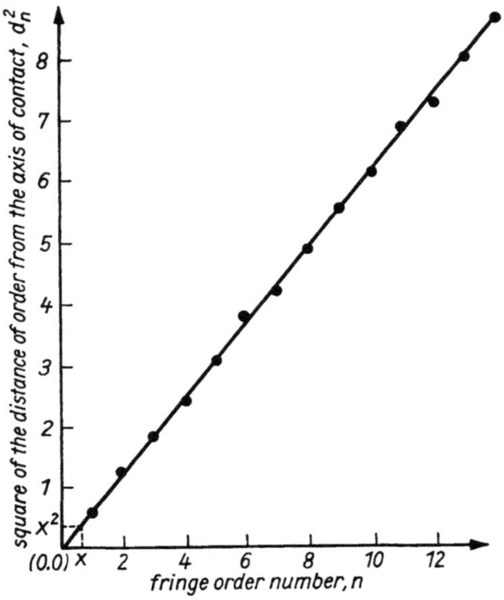

Fig. 4. Plot of the square of the distance of order from axis of contact (d_n^2) against fringe order number (n)

1. Distance between the two first order d fringes (D) $= 0.1650$ mm .
2. Distance of the first order fringe from the axis of contact $D_1 = D/2$ $= 0.0825$ mm .
3. Displacement of the first order fringe at the scratch from the axis of contact (D_1') $= 0.1435$ mm .
4. $X = D_1' - D_1 = 0.1425 - 0.0825 = 0.0600$ mm

$X^2 = 0.0036 = 0.36 \times 10^{-2}$ mm .

Table 2

Thickness parameters of various films

Materials of thin films	fractional order dn	thickness of thin film/Å
silver	0.19	518
	0.26	709
	0.33	900
	0.40	1092
	0.46	1255
aluminium	0.35	955
	0.43	1173
	0.52	1419
	0.61	1665
	0.73	1992
antimony	0.57	1556
	0.65	1774
	0.76	2074
	0.83	2265
	0.94	2566

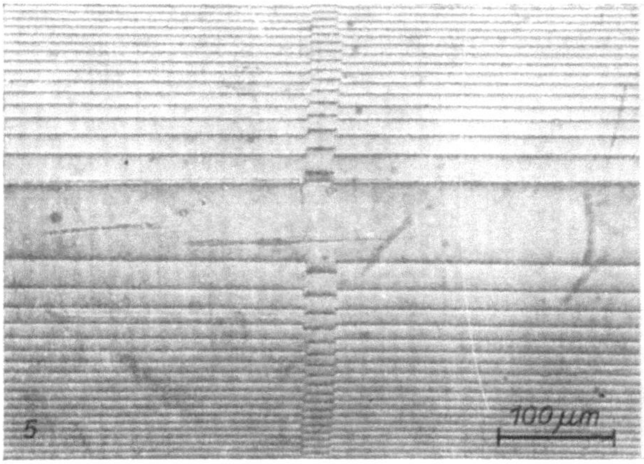

Fig. 5. Multiple beam interferogram of aluminium thin film

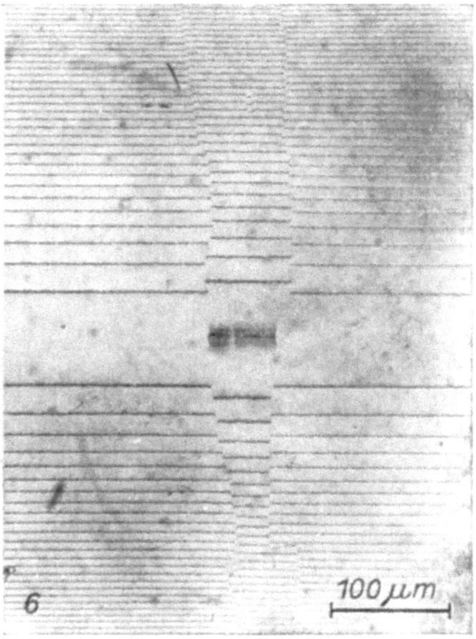

Fig. 6. Multiple beam interferogram of antimony thin film

5. For this value of X^2, the corresponding value of the fraction of order $n = 0.67$.
Actual order $dn = 1 - n = 1 - 0.67 = 0.33$.
6. Calculated value of the thickness of the thin film (t) $= dn\left(\frac{\lambda}{2}\right)$

$$= \frac{0.33 \times 5460}{2} \text{Å}$$

$$= 900 \text{ Å}.$$

Thin films of various thickness of silver, aluminium and antimony materials were prepared and by adopting the same procedure, thickness parameters of above films were determined and the important results are summarized in Table 2. Figures 5 and 6 represent clearly the multiple beam interferograms on aluminium and antimony films, respectively. It is observed that the surfaces of aluminium and antimony films are very smooth.

4. Conclusions

Advantages of this method are as follows:

1. It removes the ambiguity in allocation of fringe orders.
2. It measures the thickness parameter exactly at the place where one desires to know the thickness accurately.
3. It can be used to measure thickness of films deposited not only by thermal evaporation but by any other method.
4. Interferograms over the surface of films clearly reveals the smoothness of surfaces.
5. This technique gives high resolution with a good degree of reproducibility.

Precautions to be taken are as follows:

1. While the films are scratched, the substrate surface should not be scratched.
2. All the desired properties of the thin films to be determined first and then the thickness parameter should be measured.

The author thanks Dr. K. C. Desai, Mrs. Lilavati Desai for their keen interest in this work. The author is very grateful to University Grants Commission, New Delhi for sanction of the Project, "Studies of growth and characterization and properties of alkali halides and tartrate single crystals".

References

HEAVENS, O. S.: Optical properties of thin solid films, London 1985, pp. 105—113
TOLANSKY, S.: An Introduction to interferometry, London 1955, pp. 161—179
TOLANSKY, S.: Surface microtopography, London 1960, pp. 62—68

(Received January 26, 1987)

Authors' address:

Professor C. C. DESAI
Department of Physics, Sardar Patel University
Vallabh Vidyanagar 388120, Gujarat
India

Cryst. Res. Technol.	22	1987	8	1085–1088

N. N. SIROTA, I. D. LOMAKO

Institute of Physics of Solids and Semiconductors
Byelorussian Academy of Sciences, Minsk

On the Correlation between Anisotropy of Gyrotropic and Magnetic Properties of Samarium and Terbium Orthoferrites

Rare earth orthoferrites are interesting for investigation and promising materials for practical application. This paper deals with the existence of the correlation between the polar diagrams of polarization plane rotation obtained by us (SIROTA, LOMAKO 1976, 1978, 1979) and those of magnetic torques plotted from the data of BELOV et al. 1967, 1968a, 1968b). In this case the relationship is established between the optical anisotropy and crystallographic magnetic anisotropy of the orthoferrites crystals under investigation.

Редкоземельные ортоферриты являются интересным объектом для исследования и перспективным материалом для практического применения. В настоящей статье обращается внимание на существование связи между полученными нами полярными диаграммами углов поворота большой оси эллипса поляризации в ортоферритах самария и тербия (Сирота, Ломако, 1976, 1978, 1979) с одной стороны, и полярными диаграммами магнитных крутящих моментов, построенными нами по данным работ Белова с сотр. (Белов с сотр. 1967, 1968). При этом устанавливается связь между оптической анизотропией и кристаллографической магнитной анизотропией исследуемых кристаллов ортоферритов.

1. Introduction

Rare earth orthoferrites are known to have a distorted perovskite structure and to refer to the rhombic syngony, they are biaxial, birefringent crystals optically. In magnetic aspect the samarium and terbium orthoferrites are inequivalent: the axis of weak ferromagnetism is directed along the crystallographic axis [001] for $TbFeO_3$ at room temperature and for $SmFeO_3$ it is parallel to the a-axis [100]. The study of biaxial crystals by the polarized light for clarifying the peculiarities of anisotropic structures seems to be reasonable. The linearpolarized light transforms into elliptically polarized one while passing through the anisotropic absorbing crystal. The studies of the variation of the angles of rotation of the polarization ellipse large axis $\gamma(\varphi)$ (gyrotropic properties) as a function of the direction in single crystals allowed us to define the type of polar diagram symmetry for different crystallographic planes (SIROTA, LOMAKO 1978). The change in polar diagram of the rotation of the ellipse large axis of polarization caused by the effect of birefringence, optical activity and dichroism has been theoretically estimated (GRECHUSHNIKOV et al.). The polar diagrams of gyrotropic properties of orthoferrites of samarium and terbium are compared with polar diagrams of torque moments $M(\varphi)$ plotted in terms of the data reported by BELOV et al. (1967, 1968). A similiarity has been found in the character of change of the above mentioned polar diagrams depending on the crystallographic orientation (SIROTA, LOMAKO 1981).

The samarium and terbium orthoferrites have a perovskite structure with monoclinic distortions. $SmFeO_3$ is characterized by minimum monoclinic distortion $\beta = 92°$ (TREVES) and for $TbFeO_3$ — $\beta = 93°$. The character of the orthoferrite crystal structure distortions is affected by the change of the bonding angles Θ_1 and Θ_2. General regularities for magnetic and optical properties of rare earth orthoferrites have been established on the basis of distortions of crystal structure of the material.

2. Experimental

The $SmFeO_3$, $TbFeO_3$ single crystals, melt-grown by the spontaneous crystallization method under static conditions were investigated by us. The mixture has been calculated from the stoichiometric ratio of the rare earth oxides to the ferric oxide as 1:1. The effect of the solvent composition and the growth conditions on the size and perfection of the crystals has been reported elsewhere (LOMAKO, GES').

The angular dependences of the rotation of the large axis of polarization ellipse were measured via rotating the specimens (30 — 60 μm) in the plane normal to light propagation. A He–Ne laser with a wavelength of $\lambda = 0.63$ μm was used as a light source. Each direction of the electric vector of the incident polarized light φ was registered against the rotation angle γ with the analyzer by the compensation method (SIROTA, LOMAKO 1976, 1978).

3. Results and discussion

For samarium orthoferrite the polar diagrams of optical properties are characterized by a smooth change of $\gamma(\varphi)$ with a maximum at $\varphi \sim 45°$. For comparison the curves of the torque moments, taken in the plane (100) at room temperature for $SmFeO_3$ were replotted in the form of polar diagrams, reported by BELOV (BELOV 1968b). The strong magnetic fields ($H = 75$ kOe) were employed to obtain the solid curve (without breaks) of the torque moments for samarium orthoferrites. The easy magnetization axis at this temperature coincides with the direction [100] for $SmFeO_3$ and the hard magnetization axis lies in the plane (100). Figure 1 shows an analogy in the character of the change $\gamma(\varphi)$ as a function of the directions in the crystal, from crystallooptic measurements, and in the character of the change of the torque moments $M(\varphi)$ in the magnetic field. The shape of the polar diagrams of the $\gamma(\varphi)$ angle is determined by the crystal field structure. Besides, the change of the magnetic moment values $M(\varphi)$ against the direction in the crystal plane is determined by the peculiarities, value and crystal field orientation.

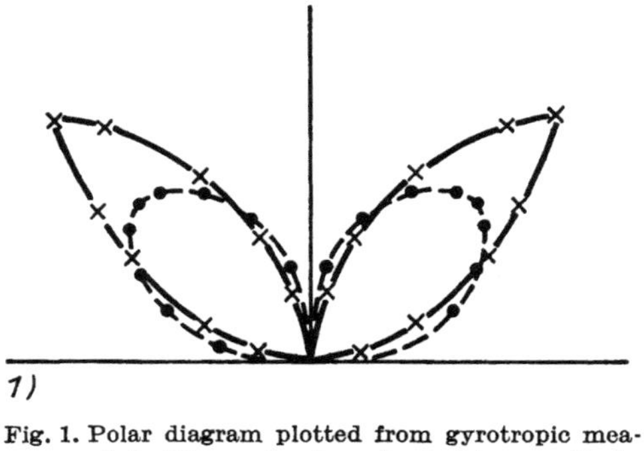

1)

Fig. 1. Polar diagram plotted from gyrotropic measurements (solid curve) and mechanical torques (dashed curve) for $SmFeO_3$, plane (100)

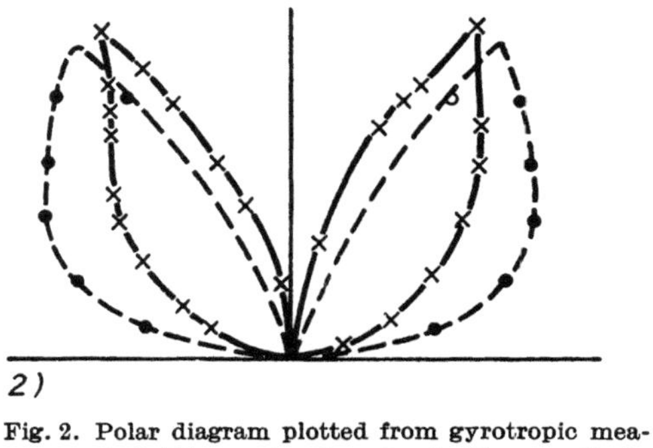

2)

Fig. 2. Polar diagram plotted from gyrotropic measurements (solid curve) and mechanical torques (dashed curve) for $TbFeO_3$, plane (110)

Figure 2 represents the part of polar diagram of the rotation angle $\gamma(\varphi)$ for terbium orthoferrite in the (110) plane plotted from our measurement data at room temperature. The value of the mechanical torque moment $M(\varphi)$ in the magnetic field (6.2 kOe) at 14 K in the same plane (110) as a function of crystallographic orientation is presented as a dashed line (BELOV et al. 1967). The replotted polar diagrams of the curves $M(\varphi)$ reveal the analogous anisotropic character with the polar diagram $\gamma(\varphi)$, with both the curves having maximum values of the above mentioned characteristics for the angle $\varphi \sim 60°$. Small discrepancy in the shape of the curves and in the directions, where the maxima $M(\varphi)$ and $\gamma(\varphi)$ are attained appears to be due to the difference in the temperatures of measurement.

While analyzing the polar diagrams obtained one can conclude, that the character of change of the rotation angle of the large axis of polarization (gyrotropic properties) coincide with that of the change of the projection of the magnetization vector in the corresponding planes.

The phenomenon of the splitting of the lobes of the polar diagram $\gamma(\varphi)$ of samarium orthoferrite observed by us at temperatures of measurement in the spin reorientation range (Fig. 3a) (SIROTA, LOMAKO 1979) is of interest. In this case the character of the polar diagram, its amplitude change essentially which is caused by different trend of temperature dependence of birefringence, dichroism and optical activity. According to the theoretical calculations the splitting of the lobes of the polar diagram $\gamma(\varphi)$ occurs for sufficiently transparent crystals, having small value of birefringence (GRECHUSHNIKOV et al.).

The analogous splitting of the polar diagram $M(\varphi)$ was observed in terms of magnetic measurements for mixed composition $Ho_{0.5}Sm_{0.5}FeO_3$, in this case no splitting was observed at small fields ($H = 6.2$ kOe) (BELOV et al. 1968b). As seen from Figure 3b the splitting was realized in the field $H = 12$ kOe. Thus the polar diagrams can be obtained by the polarization method without external effects which are analogous in their nature and form to those plotted from the magnetic measurements employing sufficiently large magnetic fields of the order of 12–75 kOe. Similar trend of the change of optical and magnetic properties of rare earth orthoferrites gives evidence of strong connection between these characteristics.

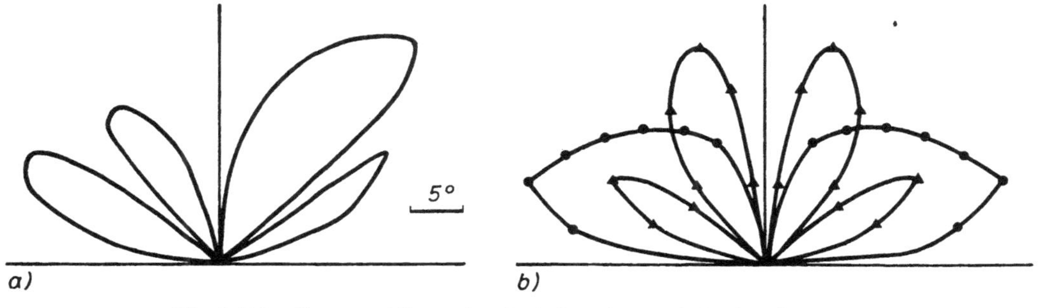

Fig. 3. Polar diagram of the angles of rotation of the polarization ellipse large axis for $SmFeO_3$, plane (100) in spin reorientation range (a); polar diagram of mechanical torque change in the magnetic field of 6.2 (●) and 12 kOe (▲) for $Ho_{0.5}Sm_{0.5}FeO_3$ (b). $T = 2.1$ K

References

BELOV, K. P., KADOMTSEVA, A. M., LEDNEVA, T. M., OVCHINNIKOVA, T. L., PONOMAREV, Ya. G., TIMOFEEVA, V. A.: FTT 9, 2790 (1967)

Belov, K. P., Kadomtseva, A. M., Levitin, R. Z., Timofeeva, V. A.: ZhETF **55**, 2155 (1968) a
Belov, K. P., Kadomtseva, A. M., Ledneva, T. M., Ovchinnikova, T. L., Timofeeva, V. A.: FTT **12**, 3758 (1968)
Grechushnikov, B. N., Konstantinova, A. F., Lomako, I. D., Kalinkina, I. N.: Kristallographiya **25** (3), 603 (1980)
Lomako, I. D., Ges', A. P.: Kinetics and Crystallization Mechanism, Minsk 1973, p. 252
Sirota, N., Lomako, I. D., Lomako, V. M.: DAN USSR **229** (5), 1105 (1976)
Sirota, N. N., Lomako, I. D.: Kristall und Technik **13** (7), 873 (1978)
Sirota, N. N., Lomako, I. D.: Kristall und Technik **14** (2), 243 (1979)
Treves, D., Eibschütz, M., Coppens, P.: Phys. Letters **18** (3), 219 (1965)

(Received January 12, 1987)

Authors' address:

Professor N. N. Sirota, Dr. I. D. Lomako
Institute of Sciences of Solids and Semiconductors, Belorussian Academy of Sciences
ul. P. Brovki 17
220 726 Minsk, USSR

Cryst. Res. Technol.	22	1987	8	1089–1093

V. RIEDE, N. SHARIF, H. NEUMANN, H. SOBOTTA, M. S. OMAR, B. R. PAMPLIN

Sektion Physik, Karl-Marx-Universität Leipzig, and
School of Physics, University of Bath

Infrared Lattice Vibrations in Cu_2GeSe_3

Infrared reflectivity spectra of Cu_2GeSe_3 are measured at room temperature in the wavenumber range from 180 to 4000 cm^{-1}. From an analysis of the spectra the parameters of four vibrational modes are determined. The experimental results are compared with predictions from group theory. From a comparison of the results for Cu_2GeSe_3 with the vibrational characteristics of other chalcogenides it follows that the force constants of cation-chalcogen bonds increase with increasing valence of the cation.

Infrarot-Reflexionsspektren von Cu_2GeSe_3 wurden bei Raumtemperatur im Wellenzahlbereich von 180 bis 4000 cm^{-1} gemessen. Durch Analyse der Spektren wurden die Parameter von vier Schwingungsmoden bestimmt. Die experimentellen Ergebnisse werden mit gruppentheoretischen Voraussagen verglichen. Aus einem Vergleich der Ergebnisse für Cu_2GeSe_3 mit den Gitterschwingungseigenschaften anderer Chalkogenide folgt, daß die Kraftkonstanten der Kation-Chalkogen-Bindungen mit steigender Wertigkeit des Kations wachsen.

1. Introduction

The ternary compound Cu_2GeSe_3 belongs to the family of diamond-like semiconductors of general composition $A_2^IB^{IV}C_3^{VI}$ with A^I = Cu or Ag, B^{IV} = Si, Ge or Sn and C^{VI} = S, Se or Te which can be derived from the binary $A^{II}B^{VI}$ compounds by cross substitution of the A^{II} atom (GOODMAN). In recent years, many of these ternary semiconductors have been synthesized, and their structural, mechanical, thermal and electrical properties have been investigated by various authors and have been reviewed by BALANEVSKAYA et al., DOVLETOV, GORYUNOVA, and by PARTHÉ, GARIN. For Cu_2GeSe_3 and several other $A_2^IB^{IV}C_3^{VI}$ compounds also optical measurements have been published, but in all cases the results are restricted to the photon energy range near the fundamental absorption edge (BALANEVSKAYA et al.; DOVLETOV). In the present communication we report the first infrared optical study of Cu_2GeSe_3. From an analysis of the optical spectra the parameters of four vibrational modes are determined. The experimental results are compared with predictions from group theory. Furthermore, an interesting trend in the interatomic force constants for cation-chalcogen bonds is found from a comparison of the vibrational characteristics of Cu_2GeSe_3 with those of other chalcogenides with tetrahedral coordination.

2. Experimental

The samples used in the measurements were parallel-sided polycrystalline specimens with typical diameters of 3–4 mm and thicknesses in the range from 0.5 to 1 mm. They were cut from larger ingots grown by the Bridgman technique using the prereacted compound as starting material. X-ray diffraction measurements confirmed the orthorhombic structure and the lattice parameters of the compound as they have been reported by PARTHÉ and GARIN. The infrared reflectivity spectra of the samples were recorded at room temperature in the wavenumber range $\bar{\nu} = 180-4000\ cm^{-1}$ with incident radiation polarized perpendicular to the plane of incidence using a Perkin-Elmer spectrophotometer model 580B. None of the samples showed a measurable transmittance.

3. Results and discussion

Although all the samples investigated were polycrystalline we observed in most cases a weak polarization dependence of the reflectivity spectra. Obviously, this means that there is some preferential orientation of the single crystal grains at the sample surface which, however, could not be detected by X-ray diffraction measurements. The two types of reflectivity spectra always obtained for polarization directions which are perpendicular to each other exhibit four and three well pronounced structures, respectively, in the wavenumber range below $350\ cm^{-1}$, which can be ascribed to optical vibrational modes. The corresponding part of the spectra is shown in Fig. 1. No further structures were found at $\bar{\nu} > 400\ cm^{-1}$.

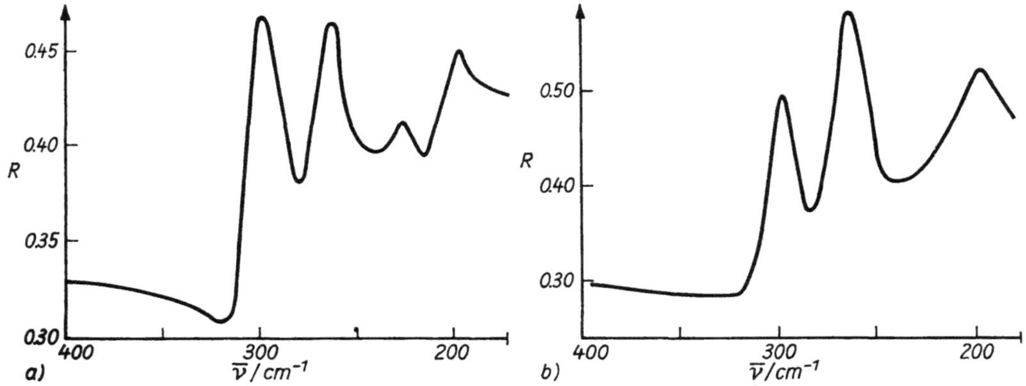

Fig. 1. Infrared reflectivity spectra of a polycrystalline Cu_2GeSe_3 sample at 300 K measured for two polarization directions which are perpendicular to each other

To determine the optical mode parameters, both types of spectra were first analysed by the Kramers-Kronig method. The mode frequencies obtained in this manner were then used as starting parameters for an analysis of the spectra in terms of a multiharmonic damped oscillator model using a complex dielectric function of the form

$$\varepsilon(\bar{\nu}) = \varepsilon_\infty + \sum_{k=1}^{n} \frac{s_k \bar{\nu}_k^2}{\bar{\nu}_k^2 - \bar{\nu}^2 - i\gamma_k \bar{\nu}_k \bar{\nu}} \,. \tag{1}$$

Here, ε_∞ is the high-frequency dielectric constant, n the number of lattice oscillators, and $\bar{\nu}_k$, s_k and γ_k are the frequency, the oscillator strength and the damping constant of the kth mode. The wavenumbers $\bar{\nu}_{TO}$ and $\bar{\nu}_{LO}$ of the transverse and longitudinal vibrational modes were determined from the maxima in the imaginary part of the dielectric function $\text{Im}\,(\varepsilon(\bar{\nu}))$ and in the energy loss function $\text{Im}\,(-1/\varepsilon(\bar{\nu}))$, respectively.

The resulting characteristic parameters of the lattice modes determined from the spectra of Figures 1a and 1b are compiled in Tables 1 and 2, respectively. The values

Table 1

Optical mode parameters determined from the reflectivity spectrum of Figure 1a

Mode	Kramers-Kronig method		oscillator model			
	$\bar{\nu}_{TO}$/cm^{-1}	$\bar{\nu}_{LO}$/cm^{-1}	$\bar{\nu}_{TO}$/cm^{-1}	$\bar{\nu}_{LO}$/cm^{-1}	s_k	γ_k
1	300	309	299	305	0.59	0.052
2	267	272	265	271	0.59	0.064
3	230	231	228	229	0.06	0.027
4	202	205	199	204	0.70	0.11

Table 2

Optical mode parameters determined from the reflectivity spectrum of Figure 1b

mode	Kramers-Kronig method		oscillator model			
	$\bar{\nu}_{TO}$/cm^{-1}	$\bar{\nu}_{LO}$/cm^{-1}	$\bar{\nu}_{TO}$/cm^{-1}	$\bar{\nu}_{LO}$/cm^{-1}	s_k	γ_k
1	298	309	298	304	0.5	0.043
2	264	275	263	278	1.3	0.055
3	203	215	202	217	1.7	0.117

for the high-frequency dielectric constant varied from sample to sample in the range $\varepsilon_\infty = 11.4-13.8$ without any clearly pronounced polarization dependence.

Comparing the results for the two polarization directions with each other we see that the mode 1 with highest energy has approximately the same parameters in both cases. Furthermore, within the error limits of about ± 3 cm^{-1} in the frequency determination the modes 2 and 4 of the spectrum of Fig. 1a have the same transverse optical mode frequencies as the modes 2 and 3 of the spectrum of Figure 1b (see Tables 1 and 2). However, the oscillator strengths of these two modes are by a factor of more than two larger for the latter polarization direction. Furthermore, the longitudinal mode frequency $\bar{\nu}_{LO}$ of mode 3 in Table 2 is distinctly larger than that of mode 4 in Table 1. Finally, mode 3 of Table 1 is clearly polarization dependent since it disappears completely in the other polarization direction (see Fig. 1b).

Now, let us compare the experimental data with the results of a group theoretical analysis of the vibrational spectrum of Cu_2GeSe_3. According to PARTHÉ and GARIN this compound crystallizes in an orthorhombic lattice with space group C_{2v}^{20} and two molecular units per unit cell. The positions of the atoms are 4 Cu in 4(c), 2 Ge in (2b), 2 Se in 2(a) and 4 Se in 4(c) (in Wyckoff notation). On the basis of this result and taking into account that the primitive cell contains only one molecular unit we have performed a group theoretical analysis of the lattice vibrations using the correlation method (FATELEY et al.). The resulting irreducible representation of the phonon normal modes at the centre of the Brillouin zone is

$$\Gamma_{ac} = 1\,A_1 + 1\,B_1 + 1\,B_2 \tag{2}$$

for the acoustic modes and

$$\Gamma_{opt} = 5\,A_1 + 2\,A_2 + 5\,B_1 + 3\,B_2 \tag{3}$$

for the optical modes. The A_1 modes are infrared active for the polarization direction $\boldsymbol{E} \parallel \boldsymbol{c}$, the B_1 modes for $\boldsymbol{E} \parallel \boldsymbol{a}$ and the B_2 modes for $\boldsymbol{E} \parallel \boldsymbol{b}$. All the optical modes are Raman active.

From (3) we see that only three infrared active modes are expected for $\boldsymbol{E} \parallel \boldsymbol{b}$. Thus, although no conclusions regarding the symmetry of the observed modes can be drawn from our experiments it is certainly justified to suppose that the polarization direction giving rise to the spectrum of Figure 1a with four vibrational bands does not correspond to $\boldsymbol{E} \parallel \boldsymbol{b}$, but to $\boldsymbol{E} \parallel \boldsymbol{a}$ or $\boldsymbol{E} \parallel \boldsymbol{c}$. Consequently, the additional mode 3 (see Tab. 1) present only in this spectrum should have A_1 or B_1 symmetry. The other three lattice oscillators observed in our experiments have approximately the same transverse mode frequencies for both polarization directions (see Tab. 1 and 2). Therefore, it cannot be excluded that vibrational modes of all the symmetries A_1, B_1 and B_2 with frequencies very close to each other and to the measured ones contribute to the corresponding bands in the reflectivity spectra. If this supposition is true the frequencies $\bar{\nu}_{TO}$ of the modes 1, 2, and 3 of Table 2 should be nearly equal to the transverse mode frequencies of the three B_2 modes predicted by group theory. In other words, there should be no modes with B_2 symmetry below 180 cm^{-1}.

Since the orthorhombic lattice of Cu_2GeSe_3 is a superstructure of the cubic sphalerite lattice (PARTHÉ, GARIN) it can be supposed that in complete analogy to the situation in the $A^IB^{III}C_2^{VI}$ chalcopyrite semiconductors (NEUMANN 1985a) the mode 1 with highest energy is sphalerite-like in nature, and their frequency should be essentially determined by the strength of the Ge–Se bond in the compound. Furthermore, this mode should be present in all polarization directions with nearly the same characteristic parameters which agrees with our experimental findings (see Tab. 1 and 2). On the basis of this assumption some conclusions can be made regarding the trends in the interatomic force constants describing the next-nearest neighbour interaction of cation-chalcogen bonds. From previous studies of the vibrational properties of the $A^{II}B^{VI}$ and $A^IB^{III}C_2^{VI}$ compounds (NEUMANN 1985a, b) we know that the frequencies of the sphalerite mode due to the A^{II}–B^{VI} bond in a $A^{II}B^{VI}$ compound are always below those of the sphalerite-like mode due to the B^{III}–C^{VI} bond in a $A^IB^{III}C_2^{VI}$ compound, if the C^{VI} atom is the same and the cations are from the same row of the Periodic Table. For instance, in ZnSe we have $\bar{\nu}_{TO} = 205\ cm^{-1}$, and $\bar{\nu}_{TO}(Ga\text{–}Se) = 250\ cm^{-1}$ in $CuGaSe_2$. From the results of the present study we have $\bar{\nu}_{TO}(Ge\text{–}Se) \approx 300\ cm^{-1}$. Thus, going from the Zn–Se to the Ga–Se and, finally, to the Ge–Se bond we have a clearly pronounced increase in the mode frequency which can only be explained by an increase in the force constants for next-nearest neighbour interaction. In other words, we observe the interesting tendency that the force constants for cation-chalcogen bonds increase with increasing valence of the cation. It is interesting to note that no such tendency seems to exist in the pnictides. For instance, we have $\bar{\nu}_{TO} = 367\ cm^{-1}$ in GaP (NEUMANN 1985b) and $\bar{\nu}_{TO} = 375\ cm^{-1}$ in $CuGe_2P_3$ (RIEDE et al. 1984), and a comparative analysis of the elastic constants of both compounds gave also no essential difference in the force constants (HAILING et al.). At present, no explanation of these trends in the force constants can be given.

Finally, some remarks regarding the high-frequency dielectric constant of Cu_2GeSe_3. According to LEVINE the value of ε_∞ in Cu_2GeSe_3 is given by

$$\varepsilon_\infty = 1 + \tfrac{1}{3}\,[2\chi(\mathrm{CuSe}) + \chi(\mathrm{GeSe})]\,, \tag{4}$$

where χ(CuSe) and χ(GeSe) are the susceptibilities of the Cu–Se and Ge–Se bonds, respectively. From the bond susceptibility model proposed by LEVINE (see also NEUMANN 1983) we find $\chi(\mathrm{GeSe}) = 9.9$. Furthermore, susceptibilities for the Cu–Se bond have been estimated from the dielectric constants of the $A^IB^{III}C_2^{VI}$ compounds with the result that $\chi(\mathrm{CuSe}) = 6.0 \pm 0.4$ (NEUMANN 1983). Inserting these two values in relation (4) we obtain $\varepsilon_\infty = 8.3 \pm 0.3$ for Cu_2GeSe_3. This value is well below those in the range $\varepsilon_\infty = 11.4-13.8$ determined from the reflectivity spectra. Obviously, this discrepancy between estimated and experimental ε_∞ values is due to the polycrystallinity of the samples. In general, if a sample shows no optical transmittance the measured reflectance is equal to the true reflectivity R of the material and can be used to estimate ε_∞ by means of the relation

$$\varepsilon_\infty = \left(\frac{1 + \sqrt{R}}{1 - \sqrt{R}}\right)^2 \tag{5}$$

in the wavenumber range well beyond the lattice vibrational modes. However, this supposition is only true for single crystals. In the case of polycrystalline material the reflectance can be higher than the reflectivity R due to internal reflections at grain boundaries even if the transmittance is very low. Obviously, this effect gives rise to, firstly, an enhanced value of ε_∞ from relation (5) and, secondly, some scatter in the ε_∞ values from sample to sample. Similar results have been obtained also in other compounds. For instance, from reflectance measurements on polycrystalline $CuInTe_2$ we found $\varepsilon_\infty = 16$ (Riede et al. 1979) while the true value of the high-frequency dielectric constant of this compound has been determined to be $\varepsilon_\infty = 9.3 \pm 0.3$ (Riede et al. 1980) which is well below the value for polycrystalline material. Therefore, the values of ε_∞ in the range from 11.4 to 13.8 found in the present study for polycrystalline Cu_2GeSe_3 can only be considered as an upper bound for the high-frequency dielectric constant of this compound.

References

Balanevskaya, A. E., Berger, L. I., Petrov, V. M.: Izv. Akad. Nauk SSSR, Neorg. Mater. **2**, 810 (1966)

Dovletov, K.: Izv. Akad. Nauk SSSR, Neorg. Mater. **19**, 1309 (1983)

Fateley, W. G., Dollish, F. R., McDevitt, N. T., Bentley, F. F.: Infrared and Raman Selection Rules for Molecular and Lattice Vibrations, New York 1972

Goodman, C. H. L.: J. Phys. Chem. Solids **6**, 305 (1958)

Goryunova, N. A.: Halbleiter mit diamantähnlicher Struktur, Leipzig 1971

Hailing, T., Saunders, G. A., Omar, M. S., Pamplin, B. R.: J. Phys. Chem. Solids **45**, 163 (1984)

Levine, B. F.: J. Chem. Phys. **59**, 1463 (1973)

Neumann, H.: Cryst. Res. Technol. **18**, 1299 (1983)

Neumann, H.: Helv. Phys. Acta **58**, 337 (1985a)

Neumann, H.: Cryst. Res. Technol. **20**, 773 (1985b)

Parthé, E., Garin, J.: Monatsh. Chem. **102**, 1197 (1971)

Riede, V., Neumann, H., Sobotta, H., Kissinger, W., Omar, M. S., Pamplin, B. R.: phys. stat. sol. (b) **123**, K119 (1984)

Riede, V., Neumann, H., Sobotta, H., Tomlinson, R. D., Elliott, E., Haworth, L.: Solid State Commun. **33**, 557 (1980)

Riede, V., Sobotta, H., Neumann, H., Nguyen, H. X., Möller, W., Kühn, G.: phys. stat. sol. (b) **93**, K93 (1979)

(Received January 30, 1987)

Authors' addresses:

Dr. V. Riede, N. Sharif, Dr. H. Neumann, Dr. H. Sobotta
Sektion Physik, Karl-Marx-Universität
Linnéstraße 5
Leipzig
DDR-7010

Dr. M. S. Omar, Dr. B. R. Pamplin
School of Physics, University of Bath
Claverton Down
Bath BA2 7AY
England

Cryst. Res. Technol.	22	1987	8	1095–1099

S. M. Dharma Prakash, P. Mohan Rao

Department of Physics, Mangalore University, Mangala Gangothri,
Mangalore, India

Microhardness Investigations on Gel-grown Barium-Cadmium Oxalate Mixed Crystals

Microhardness characteristics of barium-cadmium oxalate crystals were undertaken using the indentation technique and the results there of are reported. The observed results were compared with the microhardness values of barium oxalate dihydrate and cadmium oxalate trihydrate crystals. The indentation induced cracks etched in 1N HCl reveal the surface of the dislocation pattern arround indentation mark. The mixed crystals show higher values of microhardness than the components barium oxalate dihydrate and cadmium oxalate trihydrate crystals.

1. Introduction

Micromechanical hardness testing provides useful information concerning the mechanical behaviour of metals and minerals. Buckle has pointed out the possibility of investigating various material properties by means of microhardness measurements. Whenever hardness is of prime importance for a particular requirement, mixed crystals are used because the microhardness of mixed crystals always exceeds the microhardness of the component single crystals. Smakula et al. from their measurements of the microhardness of pottassium chloride-potassium bromide mixed crystals, have found that mixed crystals show higher hardness values than the pure components. Several investigators (Subba Rao, Haribabu; Subrahmaniam, Bansigir; Hopkins et al.) have observed similar results in various mixed crystals. Barium oxalate dihydrate and cadmium oxalate trihydrate, from mixed crystals in all proportions. A literature survey reveals that there has been no investigation on mechanical behaviour of these crystals. Hence systematic studies on microhardness characteristics of these crystals were undertaken and the results there of are reported in the present paper. Microhardness of mixed oxalate crystals has been measured using the indentation technique.

2. Experimental

Mixed crystals of barium-cadmium oxalate were grown by gel technique (Dharma Prakash, Mohan Rao) at ambient temperature using 'Analar' grade chemicals of barium chloride, cadmium chloride and oxalic acid as starting materials. Microhardness measurements were performed at room temperature using the microhardness tester Mo6270 manufactured by Vickers Ltd. U.K. on the as-grown (110) planes of the mixed oxalate crystals. The indentor orientation was fixed throughout the measurements along the ⟨100⟩ direction with respect to the plane studied. Loads varying from 20 to 100 g were applied over a fixed interval of time, say 30 s and diagonal length at different loads were determined. The diagonals of the indented impressions obtained at various applied loads were measured using the filar micrometer of the Vickers projection metallurgical microscope. About twenty indentations were made on crystal surface for each loading, the mean value of the diagonal was found. This value of diagonal of indentation was used to calculate the microhardness VHN, using the expression VHN $= 1.854 \times P/d^2$ where P is the applied load (kg) and d is the diagonal length of the indentation figures (mm). The observed results were compared with the microhardness values of barium oxalate dihydrate and cadmium oxalate trihydrate crystals. In addition, the indentation induced cracks were etched in 1N HCl solution, which is identified as one of the dislocation etchant (DHARMA PRAKASH, MOHAN RAO) to reveal the surface of the dislocation pattern around indentation marks.

2.1. Uncertainty in the measurements

Diamond pyramid hardness is nearly always measured from the observed size of the impression remaining after a loaded indentor has both penetrated and been removed from the surface. Thus the observed hardness behaviour is the summation of a number of effects involved in the material response to the indentation pressure during unloading, in the relaxation during unloading.

The probable error of the mean VHN based on 5 to 10 indentation is always within 4% of the average value. But it is observed that due to slight deviation from right angle made by the indentor axis and the crystal surface, two of the four vertices of the indentation marks, on some of the samples are not well defined, and in such cases the measurements of length of diagonals are not very accurate. To minimize the error half diagonals are measured 5 to 6 rows of indentation are made on each surface and the mean diagonals used for the calculation of Vicker hardness numeral (VHN).

3. Observations and discussion

Barium cadmium oxalate crystals when indented with load about 60 g developed cracks around corners of the indentation on a (110) plane as shown in Figure 1. Here, indentation diagonals are approximately parallel to ⟨011⟩ direction. Radial cracks formed on the edges of the Vickers indent, when the system is loaded with 80 g is shown in Figure 2. These radial cracks are usually confined to a narrow zone on the specimen surface as observed. Deformation is too severe at too high loads.

Generation of dislocations due to indentation were studied by etching (110) faces indented at a load of 60 g. Rosette patterns were obtained around the indentation on etching indented (110) faces. One such case is shown in Figure 3 (etching time 30 s). It is seen that the rosette arms are symmetric around the indentation mark and etch pit density along the rosette arms is almost uniform.

The mechanical properties of the crystals are better characterised by establishing the variation of log load against log diagonal length. Figure 4 shows the variation of the diagonal length d (cm) with load P (g). Contrary to theoretical prediction supported by the formula, VHN $= 1.854 \times P/d^2$, it was found that in the case of investigated plane, the variation was linear as expected. Consequently the validity of the law $P = ad^n$ can be established, where $n = 2$ for the investigated plane. To determine exactly the value of n, logarithms of both P and d were plotted, as shown in Figure 5, which gives a slope of 2.04.

The variation of VHN values of barium cadmium oxalate mixed crystals is shown in Figure 6 along with VHN values of barium oxalate dihydrate and cadmium oxalate trihydrate crystals. The mixed crystals show higher values of microhardness than the components of barium oxalate dihydrate (DHARMA PRAKASH, MOHAN RAO) and cadmium oxalate trihydrate. The maximum increase in the microhardness is obtained in

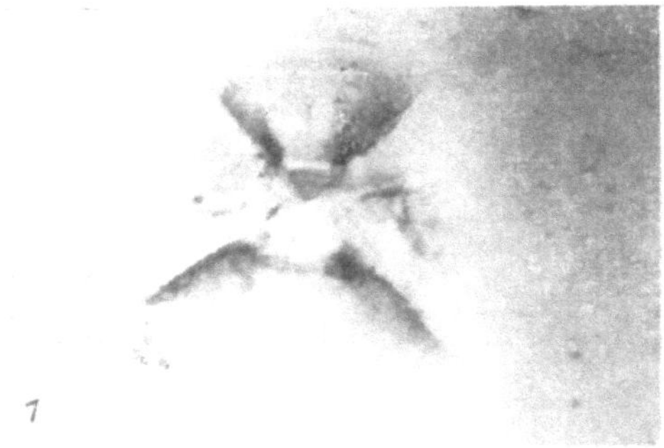

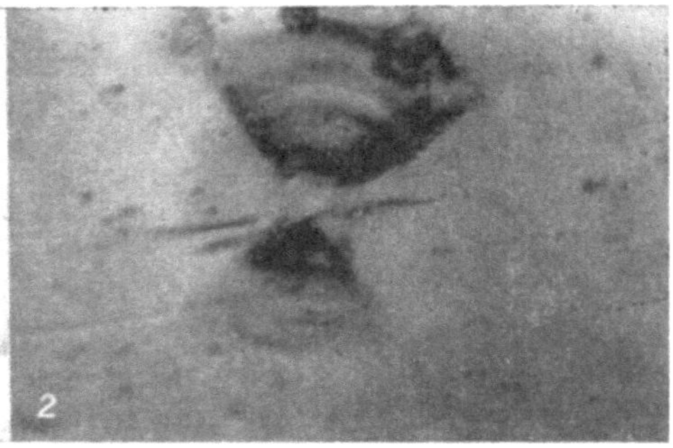

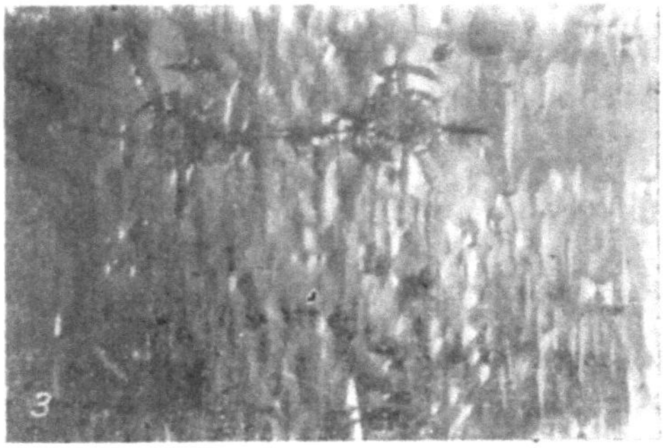

Fig. 1. The indentation mark at 60 g load on a (110) plane with the crack system

Fig. 2. Radial cracks formed on the edges of the Vickers indent, on a (110) plane at 80 g load

Fig. 3. Indentation rosettes on a (110) plane, after selective etching in 1N HCl solution (etching time 30 s)

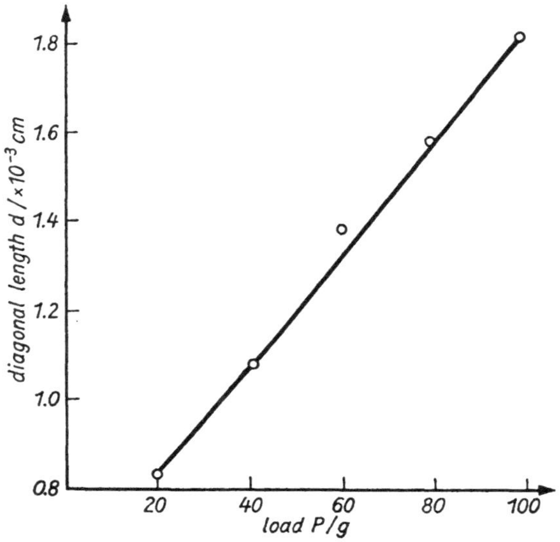

Fig. 4. Variation of the diagonal length with load

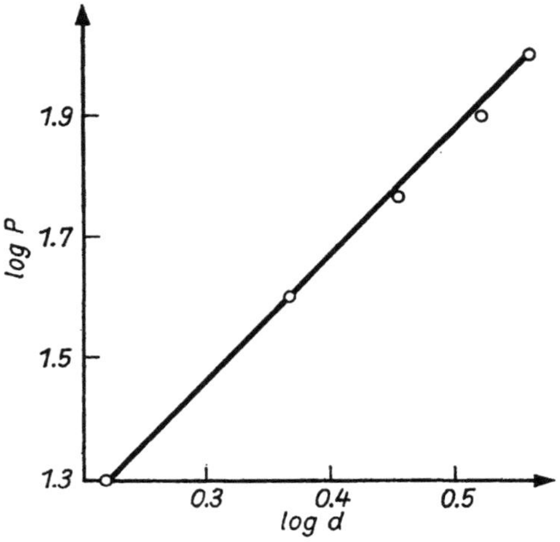

Fig. 5. Plot of log P against log d

crystals of nearly equimolar concentration (i.e., for x approximately 0.5 in $Cd_xBa_{1-x} \cdot \cdot C_2O_4$ 2.5 H_2O) and increase in the hardness is as high as 37 and 82 kg · mm^{-2} compared to the hardness of cadmium oxalate trihydrate and barium oxalate dihydrate crystals, respectively. When these results are compared with those of earlier workers it is found that the variation of microhardness for mixed crystals is similar to that observed in other mixed crystals.

Internal strains arising out of the difference in ionic sizes may be responsible for the formation of dislocations, low angle grain boundaries and other defects These defects in turn appear to be responsible for hardening in mixed crystals. The microhardness in mixed crystals depends upon the difference in the size of the ions and not in the nature of the ions substituted (SUBBA RAO, HARIBABU).

VHN for barium cadmium oxalate was found to be a function of indentor load. The average load independent hardness value is different for the same planes of barium oxalate dihydrate and cadmium oxalate trihydrate crystals studied. In the present case they are

$$\text{VHN} \ (BaC_2O_4 \cdot 2\ H_2O) = 58.27 \ \text{kg} \cdot \text{mm}^{-2}$$

$$\text{VHN} \ (CdC_2O_4 \cdot 3\ H_2O) = 103.08 \ \text{kg} \cdot \text{mm}^{-2}$$

$$\text{VHN} \ (Ba_{0.5}Cd_{0.5}C_2O_4 \cdot 2.5\ H_2O) = 140.09 \ \text{kg} \cdot \text{mm}^{-2}$$

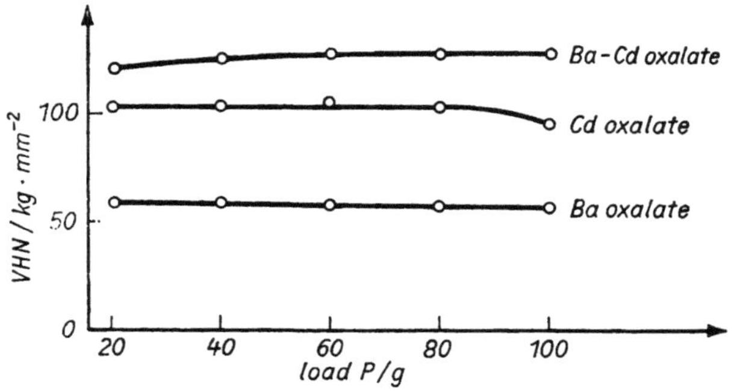

Fig. 6. Plot of Vickers microhardness number against indentation load

Obviously,

$$\text{VHN}\ Ba_{0.5}Cd_{0.5}C_2O_4 \cdot 2.5\ H_2O > \text{VHN}_{CdC_2O_4 \cdot 3H_2O} >$$
$$> \text{VHN}_{BaC_2O_4 \cdot 2H_2O}\ .$$

The interesting behaviour shown by the curves in Figure 4 becomes more understandable if we seek an explanation in terms of material chipping and the effect of the distorted zone. During the process of indentation, the indentor penetrates a depth comparable with, or greater than, the thickness of the distorted zone. Since this zone is pierced by the indentor and its effect will be marked at relatively low loads. Consequently a noticeable increase in VHN is observed in the case of barium cadmium oxalate crystals, (Fig. 6), the beginning whence the chipping of the material from the surface is very intense. There is marked decrease in the hardness value, in the case of cadmium oxalate trihydrate beyond 80 g where the chipping is not so intense. As the depth of the impression of the diamond pyramid increases, the effect of the distorted zone becomes much less prominent, which explains why the variation of VHN with load is almost nil in the case of barium oxalate dihydrate crystals. In fact, for larger loads the indentor might be reaching a depth at which distorted layers of material exist, and the microhardness ceases to be a function of load. For crystals with larger surface energies, therefore we must apply a large load to prevent the crystal surface from having a substantial effect on the microhardness.

References

Buckle, H.: Mater. Rev. **4**, 49 (1959)

Dharma Prakash, S. M., Mohan Rao, P.: J. Mater. Sci. Letts **5**, 769 (1986)

Dharma Prakash, S. M., Mohan Rao, P.: Cryst. Res. Technol. **21**, 1567 (1986)

Dharma Prakash, S. M., Mohan Rao, P.: J. Mater. Sci. Letts (communicated)

Hopkins, J. C., Miller, J. A., Martin, J. J.: phys. stat. sol. **19**, 591 (1973)

Smakula, A., Maynard, N., Repucci, A.: J. Appl. Physics **33**, 453 (1962)

Subba Rao, U. V., Hari Babu, V.: Pramana **11**, 149 (1978)

Subrahmanian, B., Bansigir, K. G.: J. Mater. Sci. **15**, 2889 (1980)

(Received February 4, 1987)

Author's address:

Dr. P. Mohan Rao, S. M. Dharma Prakash
Department of Physics, Mangalore University
Mangalagangothri-574199
Mangalore, India

Cryst. Res. Technol.	22	1987	8	1101–1105

H. Berger

Sektion Physik, Humboldt-Universität, Berlin, DDR

X-ray Determination of the Polarity of Zincblende-structure Crystals

An X-ray diffraction procedure for the determination of the polarity of noncentrosymmetric crystals by measuring the integrated reflections of opposite polarity on one surface of the crystal is described. The observed intensities must be corrected for the different asymmetries of the two reflections. To estimate the influence of crystal perfection on the correction factor, the ratios of the integrated reflections were calculated numerically for dynamic diffraction. The method has been applied for determining the polarity of CdTe.

Es wird eine Röntgenbeugungsmethode zur Bestimmung der Polarität nichtzentrosymmetrischer Kristalle durch Messung der Integralintensität an Netzebenen entgegengesetzter Polarität an einer Oberfläche des Kristalls beschrieben. Das erfordert die Berücksichtigung der verschiedenen Asymmetrien beider Reflexe. Um den Einfluß der Probenperfektion auf den Korrekturfaktor abzuschätzen, wurden die Verhältnisse der Integralintensitäten für dynamische Beugung numerisch berechnet. Das Verfahren wird zur Bestimmung der Polarität von CdTe angewendet.

1. Introduction

The polarity of surfaces of noncentrosymmetric crystals can be determined making use of the different chemical reactivity of the surfaces or of the different X-ray reflectivity of the corresponding lattice planes (Coster et al.; Bartels). The intensity of such reflections depends on the polarity because of anomalous scattering, which is strong for highly absorbing atoms. Usually, the intensities of hkl and $\bar{h}\bar{k}\bar{l}$ reflections are measured on two polar surfaces (Fewster, Whiffin). This technique demands two well prepared surfaces and remounting of the crystal. Any additional distinction of the surfaces may introduce further complications in the case of low reflectivity differences. In the following, an alternative procedure will be described. It consists of measuring two reflections of opposite polarity on the same surface and correcting the intensities for relevant geometric factors. These factors will be analyzed in detail. The reflections can be selected in such a way that the effect to be measured is large. The proposed method is applied for determining the polarity of CdTe.

2. Principle

In crystals of higher symmetry there are symmetrically nonequivalent lattice planes having the same lattice spacing. When they have opposite polarity, the corresponding reflections can be used for determining the polarity on one even surface of a large single crystal specimen. For the zincblende structure, the polarity is opposite for reflections with $h + k + l = 4n + 1$ and $h + k + l = 4n - 1$, respectively. Besides of the polarity, the asymmetries of the reflections, which are usually different, leads to different intensities. The dependence of intensity on the asymmetry must be accurately calculated. It depends also on the diffraction mechanism and, hence, on the actual crystal perfection. Therefore, both kinematic and dynamic diffraction must be considered. Because the normal scattering factors decrease with increasing $\sin\theta/\lambda$, however, the contributions of anomalous scattering are nearly independent of it, the relative influence of the anomalous scattering will be greater for higher $\sin\theta/\lambda$. Therefore, reflections with higher diffraction angles should be chosen. The measurement on one surface allows automatization and rapidly repeated measurements of the two reflections in order to eliminate intensity fluctuations and makes normalizing of the intensities unnecessary.

3. Experimental

Integrated intensities of the reflections 553 and 731 of CdTe have been measured using CuKα radiation ($\theta = 65.89°$). The specimen surfaces were etched, mechanically or chemomechanically polished. The crystal was rotated through the whole reflection region, and the detector divergence was set so large that the whole Kα spectrum and contributions due to imperfections were recorded in any case. The measurements were performed under such conditions that counting losses can be neglected and the relative standard deviations caused by counting statistics are in the order of 1%. For crystal surfaces not parallel to (111), reflections with most similar asymmetries were chosen. Care was taken in order to avoid inaccuracies due to twinning.

4. Intensity ratios for polar reflections

Using the symbols as introduced by ZACHARIASEN; COLE and STEMPLE, and BUCKSCH et al., the diffraction on polar lattice planes in the Bragg case can be described by the quantities k, p, s, and g:

$$k = |F''|/|F'| ,$$

$$p = (F'_r F''_r + F'_i F''_i)/|F'|^2 ,$$

$$s = (F'_i F''_r - F'_r F''_i)/|F'|^2 ,$$

$$g = -[(1 + b)/\sqrt{b}]\,(mc^2/e^2)\,[\mu V/(4\lambda K\,|F'|)] ,$$

where F', F'' are the structure factors calculated for real and imaginary atomic scattering factors, respectively, the indices r and i refer to real and imaginary components, μ is the linear absorption coefficient, $K = 1$ or $|\cos 2\theta|$ for σ and π polarization, $b = \sin(\theta + \varphi)/\sin(\theta - \varphi)$, φ is the angle between surface and lattice planes, further symbols have their usual meanings.

Reflections of opposite polarity differ only in the sign of s. The ratio of the $|F|^2$ of reflections of opposite polarity, symbolized by the indices H and $\bar{H}$, is

$$|F_H|^2/|F_{\bar{H}}|^2 = (1 + k^2 + 2\,s)/(1 + k^2 - 2\,s) ,$$

For mosaic crystals, the ratio of integrated reflections of opposite polarity (RIROP) as given by the kinematic theory equals

$$(I_H/I_{\bar{H}})_{\text{kin}} = (|F_H|^2/|F_{\bar{H}}|^2)\,[(1 + b_{\bar{H}})/(1 + b_H)] .$$

For perfect crystals, the dynamic theory gives for the diffraction pattern (BUCKSCH et al.)

$$R = M(1 + k^2 + 2s)/\sqrt{(1 - k^2)^2 + 4p^2}$$

with

$$M = L - \sqrt{L^2 - 1},$$

and

$$L = [1/\sqrt{(1 - k^2)^2 + 4p^2}]\,[y^2 + g^2 + \sqrt{(y^2 - g^2 - 1 + k^2)^2 + 4(gy - p)^2}],$$

where y is an angle parameter.

The integrated reflection in y units for unpolarized radiation

$$I_{\text{dyn}} = \int_{-\infty}^{\infty} R^\sigma\, dy + |\cos 2\theta| \int_{-\infty}^{\infty} R^\pi\, dy$$

can be only calculated numerically. For the RIROP in the dynamic case it follows

$$(I_H/I_{\overline{H}})_{\text{dyn}} = \sqrt{b_{\overline{H}}/b_H}\,\left(\int R_H^\sigma\, dy + |\cos 2\theta| \int R_H^\pi\, dy\right)/\left(\int R_{\overline{H}}^\sigma\, dy + |\cos 2\theta| \int R_{\overline{H}}^\pi\, dy\right) =$$
$$= (|F_H|^2/|F_{\overline{H}}|^2)\sqrt{b_{\overline{H}}/b_H}\,\left(\int M_H^\sigma\, dy + |\cos 2\theta| \int M_H^\pi\, dy\right)/\left(\int M_{\overline{H}}^\sigma\, dy + |\cos 2\theta| \int M_{\overline{H}}^\pi\, dy\right).$$

The additional factor $\sqrt{b_{\overline{H}}/b_H}$ yields from the asymmetry dependence of y.

Hence, the RIROP can be written generally as

$$I_H/I_{\overline{H}} = (|F_H|^2/|F_{\overline{H}}|^2)\, B,$$

B being the asymmetry ratio. It follows immediately that for equal asymmetry for the two reflections the RIROP's for kinematic and dynamic diffractions are identical. ($B = 1$). For $B \neq 1$ they are different.

5. Results and discussion

Because the perfection of crystals to be studied may be different and is not known exactly, there is an uncertainty in comparing experimental and calculated RIROP's. To estimate the differences between the two cases the asymmetry ratios have been calculated as functions of φ_H and $\varphi_{\overline{H}}$ for the 533/731 reflections of CdTe, CuKα radiation (Fig. 1). All the scattering factors were taken from the International Tables. As shown in Figure 1, in this instance the distinction for the two diffraction mechanisms is fairly small as far as the asymmetries are not extremely different. This results from the relatively high absorption and diffraction angles which diminish the differences between kinematic and dynamic diffraction (HIRSCH, RAMACHANDRAN). Hence, the determination of polarity can be made in this case without consideration of the perfection state of the crystal surface. Corresponding calculations for GaAs, reflection pair 551/333 lead to a greater uncertainty (for surface (111): $B_{\text{kin}} = 1.131$; $B_{\text{dyn}} = 1.115$; $|F_H|^2/|F_{\overline{H}}|^2 = 1.015$). In any case the B value must be known exactly.

Some examples of observed and calculated RIROP's considering the experimental geometric conditions are given in Table 1. In all instances the unambiguous determination of polarity is possible. The agreement between calculated and observed RIROP's is slightly better for the kinematic case. Besides that only the mean ratios of observed and calculated kinematic RIROP's (for 25 values on A faces: 0.978 $\pm$ 0.014 (dynamic), 0.987 $\pm$ 0.014 (kinematic); for 14 values on B faces: 0.974 $\pm$

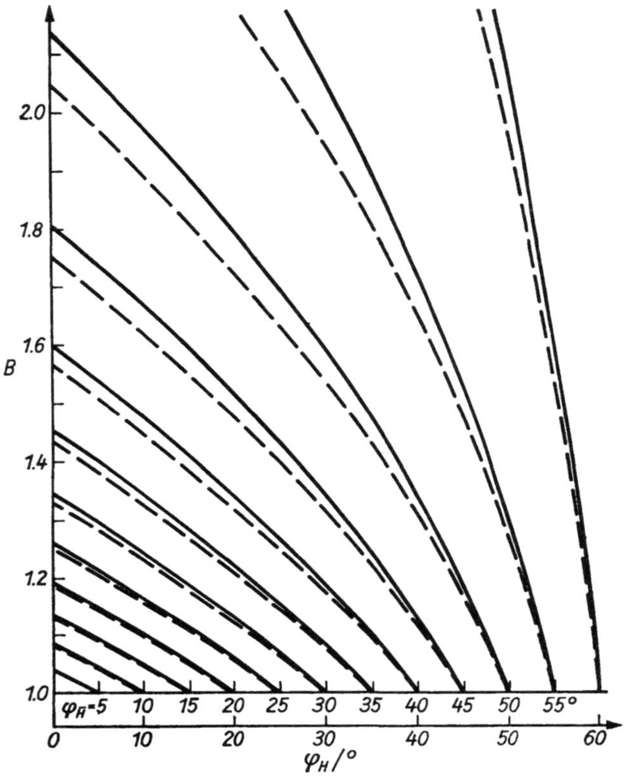

Fig. 1. Asymmetry ratios B for reflections of opposite polarity in dependence on the angles φ_H and $\varphi_{\overline{H}}$ ($\varphi_H \leqq \varphi_{\overline{H}}$) between surface and lattice planes. CdTe, CuKα, H: 553, $\overline{H}$: 731; solid line: for kinematic diffraction, dashed line: for dynamic diffraction

Table 1
Observed and calculated RIROP's and their ratios. CdTe, CuKα, H: 553, $\overline{\text{H}}$: 731 (for A(111)), H: $\bar{5}\bar{5}\bar{3}$, $\overline{\text{H}}$: $\bar{7}\bar{3}\bar{1}$ (for B($\bar{1}\bar{1}\bar{1}$))

No.	$\lvert F_H\rvert^2/\lvert F_{\overline{H}}\rvert^2$	$I_H/I_{\overline{H}}$ calc.				$I_H/I_{\overline{H}}$ obs.	A/B	$\frac{(I_H/I_{\overline{H}})_{\text{calc.}}}{(I_H/I_{\overline{H}})_{\text{obs.}}}$ for A or B, resp.	
		A		B					
		kin.	dyn.	kin.	dyn.			kin.	dyn.
1		1.057	1.060	0.862	0.864	0.894	B	0.964	0.966
						0.884		0.969	0.971
2		1.162	1.160	0.948	0.946	0.926	B	1.023	1.022
						0.917		1.033	1.031
3	1.107	1.437	1.420	1.172	1.158	1.424	A	1.009	0.997
						1.457		0.986	0.974
4		1.480	1.459	1.207	1.190	1.237	B	0.975	0.961
						1.232		0.979	0.965
5		1.376	1.362	1.122	1.111	1.406	A	0.978	0.968
						1.393		0.987	0.977

± 0.018 (dynamic), 0.982 ± 0.018 (kinematic)) show no significant deviation from unity. The spread of the values may be influenced by twinning on a microscale as well as by surface roughness.

References

Bartels, W. J.: J. Vac. Sci. Technol. **B 1**, 338 (1983)
Bucksch, R., Otto, J., Renninger, M.: Acta Cryst. **23**, 507 (1967)
Cole, H., Stemple, N. R.: J. Appl. Physics **33**, 2227 (1962)
Coster, D., Knol, K. S., Prins, J. A.: Z. Physik **63**, 345 (1930)
Fewster, P. F., Whiffin, P. A. C.: J. Appl. Physics **54**, 4668 (1983)
Hirsch, P. B., Ramachandran, G. N.: Acta Cryst. **3**, 187 (1950)
International Tables for X-ray Crystallography, Vol. IV, Birmingham 1974
Zachariasen, W. H.: Theory of X-ray Diffraction in Crystals, New York, London 1945

(Received January 28, 1987)

Author's address:

Dr. sc. H. Berger
Sektion Physik der Humboldt-Universität zu Berlin, Bereich Kristallographie
Invalidenstraße 110
Berlin
DDR-1040

Cryst. Res. Technol.	22	1987	8	1107–1113

Waldemar WOŁCZYŃSKI, Artur N. KOBYLKIN

Institute for Metal Research, Polish Academy of Sciences, Kraków, Poland, and
Institute of Metallurgy, Academy of Sciences of the USSR, Moscow

Thermal Analysis of the Microsegregation Phenomenon in Some Alloys from the Sb—Sn System

Thermal analysis in experimental verification to theoretical prediction of the maximum microsegregation phenomenon in the peritectic alloys solidification has been applied. Satisfactory agreement has been obtained between some results of theoretical calculations and experimental ones. Usibility of the DTA-5 apparatus (original construction of the Baykov Institute of Metallurgy) to detect quite small amounts of low-solidifying non-equilibrium phases has been proved.

1. Introduction

The phenomenon of the maximum microsegregation of the Sb–30 Sn alloy was described (WOŁCZYŃSKI) by means of the SCHEIL equation transformed by KRUPKOWSKI to a form as:

$$d \ln (1 - x_f) = (1 - k)^{-1} d(- \ln N_2^x) . \quad (1)$$

Additionally, the supplementary relationship, expressing the summarized effect due to nonequilibrium solidification of alloys from the peritectic phase systems was applied. It was proved theoretically that not only the (Sb)-phase but some precipitations of the $(\beta + \beta')$-phase as well as the (Sn)-phase are to be expected during solidification of the above alloy.

According to the results presented in the mentioned paper the amount of the (Sb)-phase is equal to; $x_{(\mathrm{Sb})} = 0.4585$ [mole] (with regard to 1 mole of the Sb–30 Sn alloy). Using the equation (1), the amount of each low-solidifying phase has been determined, and these quantities are equal to; $x_\beta = 0.4310$ [mole], $x_{\beta'} = 0.0366$ [mole], $x_{(\mathrm{Sn})} = 0.0739$ [mole].

It seemed possible to verify experimentally the above predictions, shown in the Figure 1, and to detect the (Sn)-phase during solidification of the Sb–30 Sn alloy. The (Sn)-phase is completely the nonequilibrium phase, and it does not exist if diffusion occurs in the solid.

It is well known (KATTAMIS, FLEMINGS; KRUPKOWSKI) that the maximum inhomogeneity is to be expected if two assumptions:

1 — diffusion in the liquid is complete,
2 — diffusion in the solid is negligible,

are valid during solidification at certain cooling rate, v_c, otherwise, at certain rate of solidification v_{heq}. The assumed verification is to be managed if this cooling rate could be found experimentally, as it was done by CIACH and DUKIET-ZAWADZKA by means of the simplified measurements of the average cooling rate.

In order to carry out the proposed experiment the DTA-5 apparatus fully tested, then described by MASLENKOW et al. was used. This apparatus allowed to cover the

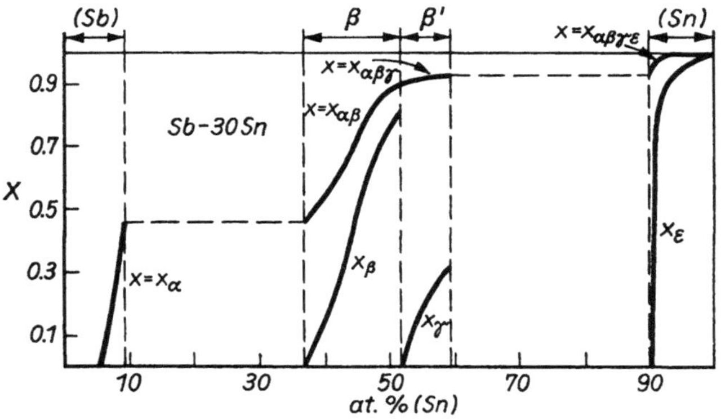

Fig. 1. Current amount of the growing dendrite vs. solid composition of the Sb–30 Sn alloy (WOŁCZYŃSKI)

following range of cooling rates: 5, 10, 25, 50, 125 and 200 [K/min]. However, it is to be noticed that above values are only the initial values of the cooling rates, and they decrease slowly during process under investigation as it was proved by MASLENKOV et al.)

2. Experiments and results

A sample of the Sb–30 Sn alloy was installed into the DTA-5 apparatus then melted and homogenized. Some solidification processes under investigation were carried out cyclically (melting — solidification), at the rate of 25 K/min for each melting process, and at the rates of 5, 10, 25, 125, 200 K/min for solidification processes, in sequence. Those experiments were thought to find the proper cooling rate (or range of cooling rates) at which some thermal effects of each phase; (Sb), β, β', (Sn), (Fig. 2) are to be detected.

The results of this verification are presented in the Figure 3, as well as in the Figures 4 and 5.

In order to confirm a correctness of the thermal effects recording an additional examine for pure components has been carried out (Fig. 4), (horizontal arrows are to point out the direction of a given process).

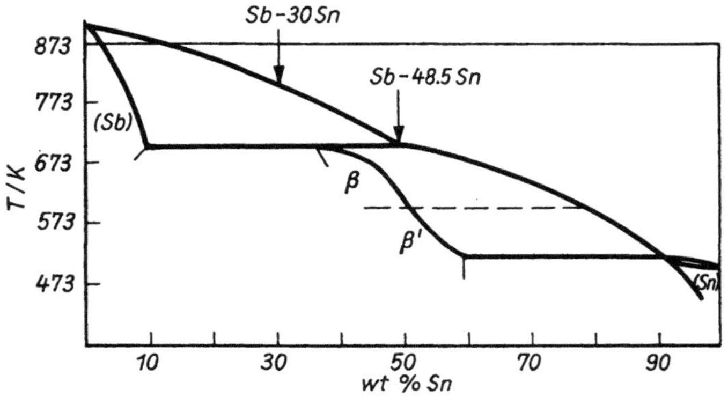

Fig. 2. The Sb–Sn phase diagram (Metals Handbook)

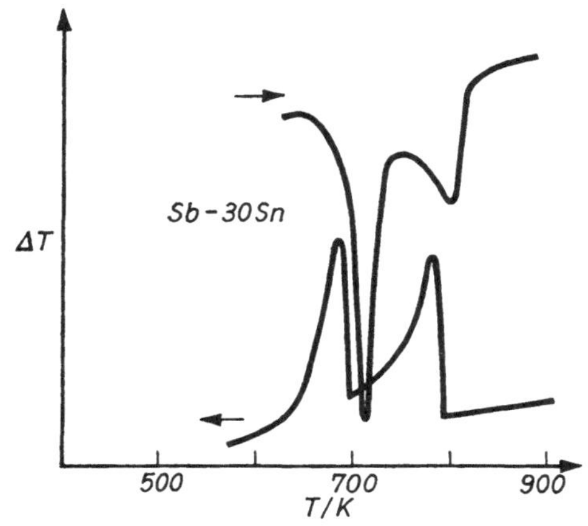

Fig. 3. The DTA-5 typical differential curves of the melting-solidification process carried out for the Sb–30 Sn alloy (heating rate of 25 K/min and cooling rate of 25 K/min

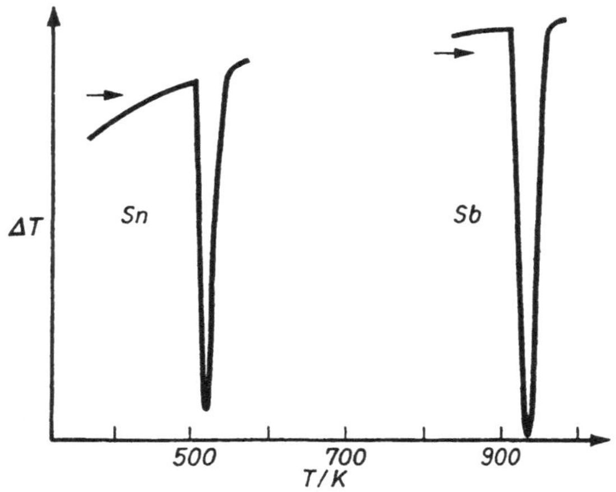

Fig. 4. The DTA-5 differential curves of the melting process carried out for the Sb and for the Sn pure components of investigated alloy, respectively

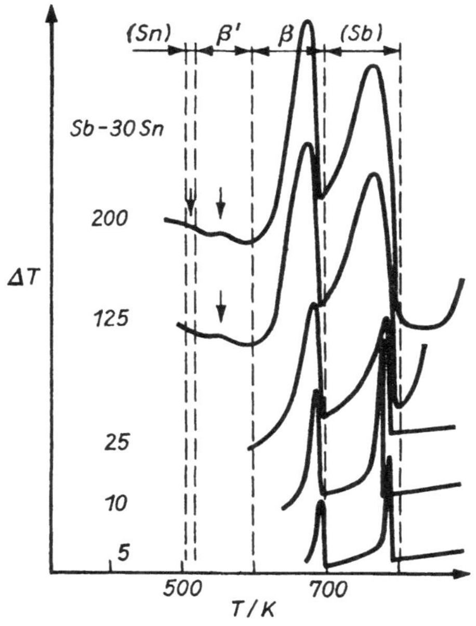

Fig. 5. Thermal effects of solidifying phases for the Sb–30 Sn alloy, which was cyclically (melting-solidification) investigated at following cooling rates; 5, 10, 25, 125, 200 K/min, in sequence

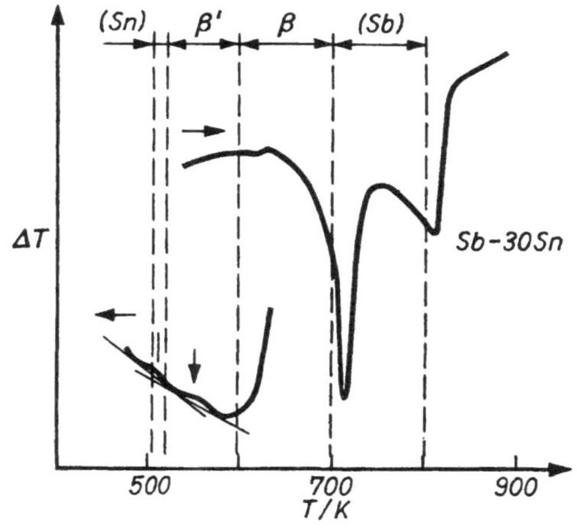

Fig. 6. Thermal effects of the Sb–30 Sn alloy solidification process carried out at heating rate of 25 K/min (without intensification of signal) and at cooling rate of 200 K/min (with intensification)

An interpretation of the obtained results has been done by use of the Sb–Sn phase diagram (Fig. 2). Therefore, in the next figures the equilibrium temperature ranges of solidification for each phase are ploted there.

Some results presented in the Figure 5 point out that:

— there are no thermal effects due to the (Sn)-phase and β'-phase formation during the Sb–30 Sn alloy solidification carried out at the cooling rates of 5, 10, 25 K/min,
— there are small thermal effects of the β'-phase formation during the Sb–30 Sn alloy solidification carried out at the cooling rates of 125 and 200 K/min, (vertical arrows),
— there is quite small thermal effect of the (Sn)-phase formation during the Sb–30 Sn alloy solidification carried out at the cooling rate of 200 K/min (vertical arrow).

As both thermal effects of the β'-phase and the (Sn)-phase formation (Fig. 5) were so small in comparison with the (Sb)-phase and the β-phase formation thermal effects, the last melting-solidification process (at cooling rate of 200 K/min) has been reproduced, with the intensification of the recorder signal for the β'-phase and the (Sn)-phase formation, only (Fig. 6).

Because of the intensification of the recorder signal the (Sb)-phase and the β-phase formation could be partly recorded, only. All the same, the thermal effects of the β'-phase and the (Sn)-phase formation are remarkable.

The same intensification of the recorder signal for the solidification at the cooling rate of 125 K/min did not lead to appearance of the remarkable thermal effect of the (Sn)-phase formation. Moreover, interpretation of the obtained differential curve was difficult because of the background noise.

For that reason a new alloy containing 48.5 weight per cent of tin was prepared to an additional investigation. Its composition was chosen in such a manner to augment the thermal effect due to the (Sn)-phase formation.

At first, the high-purity components were used to make the Sb–48,5 Sn alloy which was solidified at the cooling rate of 200 K/min. The sample of the above alloy

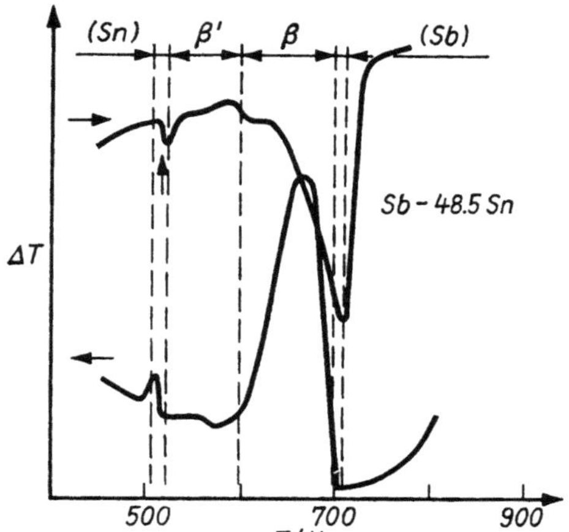

Fig. 7. The DTA-5 differential curves of the melting-solidification process carried out for the Sb–48.5 Sn alloy (heating rate of 25 K/min and cooling rate of 50 K/min)

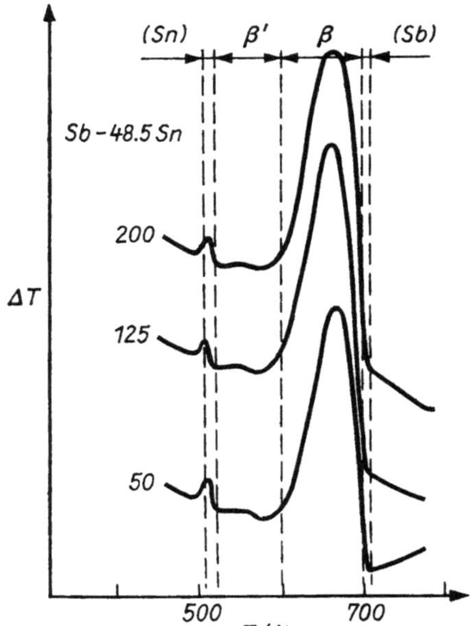

Fig. 8. Thermal effects of solidifying phases for the Sb–48.5 Sn alloy, which was cyclically (melting-solidification) investigated at following cooling rates; 200, 125, 50 K/min, in sequence

was remelted at the rate of 25 K/min then solidified at the cooling rate of 50 K/min, (Fig. 7). The thermal effect of the (Sn)-phase, (precipitated during preparation of the Sb–48.5 Sn alloy) has been recorded during its melting process (Fig. 7), (vertical arrow).

Finally, the sample of the Sb–48.5 Sn alloy was investigated cyclically (melting-solidification) using the DTA-5 apparatus to study the thermal effects of solidifying phases at the cooling rates of 200, 125, 50 K/min, in sequence, (Fig. 8).

The results presented in the Figure 8 point out that, there are some considerable thermal effects due to the β, β', (Sn)-phases formation in the Sb–48.5 Sn alloy solidification at each applied cooling rate.

The helium protective atmosphere was introduced in each of melting-solidification process. Additionally, an investigation of the empty crucible was also carried out. It was proved, that use of the crucible did not effect on the quantity as well as the quality of the recorded thermal effects.

3. Discussion

Theoretical calculations to the maximum microsegregation in solidification of the both considered alloys point out that for the v_{heq}-rate, which can assure the validity of the assumed theory, the ratios of the one phase amount to the others can be presented as follows:

a) for the Sb–30 Sn alloy solidification process;

$$x_{(\text{Sb})}/x_{\beta}/x_{\beta'}/x_{(\text{Sn})} = 1.06/1/0.07/0.17 \tag{2a}$$

b) for the Sb–48.5 Sn alloy solidification process;

$$x_{(\text{Sb})}/x_{\beta}/x_{\beta'}/x_{(\text{Sn})} = 0.04/1/0.08/0.17\,. \tag{2b}$$

The above proportion (2b) has been found on the basis of the following results of calculations: $x_{(\text{Sb})} = 0.0334$ [mole], $x_{\beta} = 0.7694$ [mole], $x_{\beta'} = 0.0653$ [mole], $x_{(\text{Sn})} = 0.1319$ [mole], (with regard to 1 mole of the Sb–48.5 Sn alloy).

Analogously, the similar ratios can be found taking into account theoretical values of heat of solidification for each considered phase:

a) of 1 mole of the Sb–30 Sn alloy,

$$L_{(\text{Sb})}/L_{\beta}/L_{\beta'}/L_{(\text{Sn})} = 1.26/1/0.07/0.09 \tag{3a}$$

b) of 1 mole of the Sb–48.5 Sn alloy,

$$L_{(\text{Sb})}/L_{\beta}/L_{\beta'}/L_{(\text{Sn})} = 0.05/1/0.07/0.09\,. \tag{3b}$$

Some areas under differential curves obtained by means of the thermal analyser are in proper proportion to heats of solidifcation for the adequate phases. Hence, it seemed possible to calculate the amount of each phase, which was formed during nonequilibrium solidification of examined alloys.

In the case of solidification of the Sb–30 Sn alloy at the rate of 200 K/min (Fig. 5) it can be measured that:

$$L_{(\text{Sb})}^{\text{exp}}/L_{\beta}^{\text{exp}}/L_{\beta'}^{\text{exp}}/L_{(\text{Sn})}^{\text{exp}} = 1.36/1/0.04/0.02\,. \tag{4}$$

Combining (2a), (3a), (4) it has been determined that in the above solidification process carried out in the DTA-5 apparatus the ratios of the one phase amount to the others are as follows:

$$x_{(\text{Sb})}^{\text{exp}}/x_{\beta}^{\text{exp}}/x_{\beta'}^{\text{exp}}/x_{(\text{Sn})}^{\text{exp}} = 1.14/1/0.04/0.04\,. \tag{5}$$

The more closed are the theoretical ratios (2a) to the experimental ones (5), the better are fulfilled the assumptions of applied model of the microsegregation phenomenon.

Otherwise, as the ratios (2a) and (5) are not identical it may be concluded that applied solidification rate was only similar to the v_{heq}-rate.

The same interpretation of the microsegregation phenomenon has been done for the Sb–48.5 Sn alloy solidification process;

a) carried out at the cooling rate of 200 K/min

$$L_{(\mathrm{Sb})}^{\exp}/L_{\beta}^{\exp}/L_{\beta'}^{\exp}/L_{(\mathrm{Sn})}^{\exp} = 0.05/1/0.07/0.084 \tag{6a}$$

and

$$x_{(\mathrm{Sb})}^{\exp}/x_{\beta}^{\exp}/x_{\beta'}^{\exp}/x_{(\mathrm{Sn})}^{\exp} = 0.04/1/0.08/0.159 \tag{7a}$$

b) carried out at the cooling rate of 125 K/min

$$L_{(\mathrm{Sb})}^{\exp}/L_{\beta}^{\exp}/L_{\beta'}^{\exp}/L_{(\mathrm{Sn})}^{\exp} = 0.05/1/0.07/0.078 \tag{6b}$$

and

$$x_{(\mathrm{Sb})}^{\exp}/x_{\beta}^{\exp}/x_{\beta'}^{\exp}/x_{(\mathrm{Sn})}^{\exp} = 0.04/1/0.08/0.147 \tag{7b}$$

c) carried out at the cooling rate of 50 K/min

$$L_{(\mathrm{Sb})}^{\exp}/L_{\beta}^{\exp}/L_{\beta'}^{\exp}/L_{(\mathrm{Sn})}^{\exp} = 0.05/1/0.07/0.072 \tag{6c}$$

and

$$x_{(\mathrm{Sb})}^{\exp}/x_{\beta}^{\exp}/x_{\beta'}^{\exp}/x_{(\mathrm{Sn})}^{\exp} = 0.04/1/0.08/0.136\,. \tag{7c}$$

As the thermal effects due to the (Sb)-phase and β-phase formation were not separated one from the other, it was assumed that areas under the β-phase differential curves in the Figure 8 may be divided in proper proportion to determine the ratios of heats of solidification for the above phases.
This assumption has been applied in the Eqs. (6).

All four phases: (Sb), β, β', and (Sn)/their amounts predicted theoretically by means of the Eq. (1)/were detected during:

a) the Sb–30 Sn alloy solidification carried out at the cooling rate of 200 K/min,
b) the Sb–48.5 Sn alloy solidification carried out at the rates of 200, 125, and 50 K/min only.

An increase of thermal effects of the all low-solidifying phases with increase of the cooling rate is remarkable (Fig. 5). Moreover, all the peaks (Fig. 5) become less sharp and transform mildly one into the other. It proves, that solidification process under investigation becomes more and more nonequilibrium one. Otherwise, it carries out more and more continuously and leads up to formation of some areas of the pure tin concentration.

4. Conclusions

Application of the thermal analysis to investigations of the microsegregation phenomenon has been successfully tested by means of the DTA-5 apparatus.

The DTA-5 apparatus has proved to be useful to detect the quite small nonequilibrium precipitations of low-solidifying phases.

The high was the cooling rate of investigated alloys the more non-equilibrium became solidification process, and some low-solidifying phases could be detected.

In the case of the Sb–30 Sn alloy solidification carried out at the cooling rate of 200 K/min the rate of this process was similar approximately to the v_{heq}-rate.

In the case of the Sb–48.5 Sn alloy solidification carried out at the three various cooling rates all these rates were closed to the v_{heq}-range of rates.

Notation

- k distribution coefficient
- L_f heat of solidification calculated for each phase f, solidifying from 1 mole of a given alloy, (f = (Sb), β, β', (Sn))
- L_f^{exp} heat of solidification (for each phase f), measured experimentally during real solidification process, then regarded to 1 mole of a given alloy, (f = (Sb), β, β', (Sn))
- N_2^x current concentration of the second component in the liquid
- T temperature, [K].
- v solidification rate, [cm³/s], [g/s]
- v_c cooling rate, [K/s]
- v_{heq} solidification rate, which can assure the validity of both assumptions concerning diffusion phenomenon
- x_f theoretical amount of each phase f, (f = (Sb), β, β', (Sn)), for solidification process carried out at the v_{heq}-rate, (with regard to 1 mole of a given alloy)
- x_f^{exp} experimental amount of each phase f, (f = (Sb), β, β', (Sn)), for real solidification process (with regard to 1 mole of a given alloy)

All the experiments of this work were carried out at the Baykov Institute of Metallurgy of the Academy of Sciences of the U.S.S.R., Moscow, under the supervision of Professor S. B. Maslenkov.

The authors thank Dr. W. Kapturkiewicz from the Institute of Foundry Technology of the University of Mining and Metallurgy, Cracow, for remarks and discussions.

References

Ciach, R., Dukiet-Zawadzka, B.: Arch. Hutn. **28**, 499 (1983)
Kattamis, T. Z., Flemings, M. C.: Trans. AIME. **233**, 993 (1965)
Krupkowski, A.: Bull. Acad. Pol. Sci. **10**, 824 (1968)
Malsenkov, S. B., Kobylkin, A. N., Sitnikov, D. N.: J. Thermal Anal. **29**, 1287 (1984)
Metals Handbook, ASM Ed. 8, 333 (1973)
Scheil, E.: Z. Metallkde. **34**, 70 (1942)
Wołczyński, W.: Cryst. Res. Technol. **21**, 1023 (1986)

(Received January 29, 1987)

Authors' addresses:

Dr. Ing. A. N. Kobylkin
Baykov Institute of Metallurgy, Academy of Sciences of the U.S.S.R., Moscow
Lenina Street 49
U.S.S.R.-W-334 Moscow

Dr. Ing. W. Wołczyński
Aleksander Krupkowski Institute for Metal Research, Polish Academy of Sciences, Kraków
Reymonta 25
Poland-30059 Kraków

Crystal Research and Technology

Zeitschrift für experimentelle und technische Kristallographie

Edited by H. Neels, Leipzig

Papers in Number 9, 1987

Original Papers

AKADEMIE-VERLAG · BERLIN

Cryst. Res. Technol.	22	1987	8	K125–K128

Short Notes

G. PELZL, M. NOVAK, W. WEISSFLOG, D. DEMUS

Forschungsabteilung der Sektion Chemie, Martin-Luther-Universität Halle-Wittenberg, DDR, and VEB Laborchemie Apolda, Betriebsteil Spezialchemie Leipzig, DDR

Binary Systems with Induced Smectic Phases Due to EDA Complex Formation

1. Introduction

Electron donor-acceptor (EDA) interaction in binary systems of liquid crystals strongly favours induced smectic phases, usually S_A phases (SHARMA et al.; DEMUS et al.; ARAYA, MATSUNAGA). Only, if the molecular structure of the mixing components leads to a special steric situation the induction of smectic phases can be suppressed in spite of strong EDA interaction (WEISSFLOG et al.). Recently by DE JEU et al. a theoretical model of induced smectic phases was developed on the base of EDA complexing.

In this paper we present two impressive examples of the induction of smectic phases due to EDA complex formation.

2. Materials

A: 4-[4-n-hexyloxybenzoyloxy]-benzylidene-4-cyanoaniline

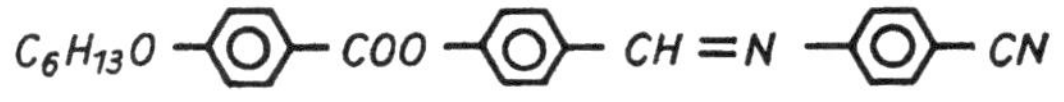

K 115 (S_A 87) N 271 Is (WEISSFLOG, PELZL et al.)

B: 4-[4-n-octyloxybenzoyloxy]-benzylidene-4-[diethylaniline]

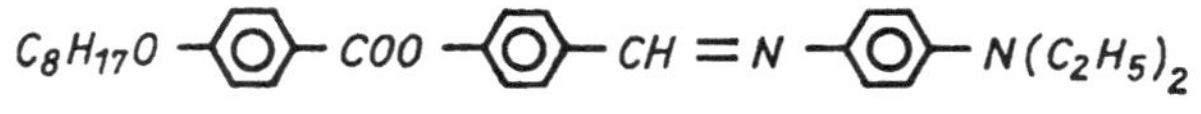

K 106 N 141.5 Is

C: 4-[4-n-octyloxybenzoyloxy]-benzylidene-4-[N-morpholino]-aniline

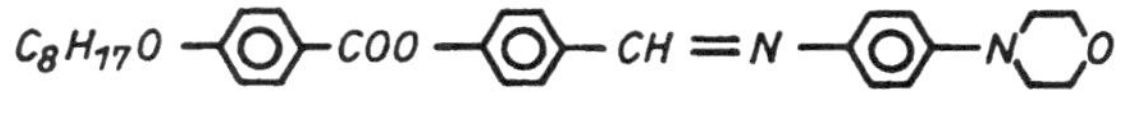

K 133 N 236 Is

The electron donor compounds B and C possess only a nematic phase, whereas the acceptor compound A forms an additional metastable S_A phase.

3. Phase Diagrams

We studied the isobaric phase diagrams of the binary mixtures A/B and A/C, respectively. The phase diagrams were investigated by the contact method (Kofler, Kofler) and by determining of transition temperatures for mixtures of different concentrations. As Figures 1 and 2 display, the most evident feature of both diagrams is an induced S_A phase in an unusual wide temperature-concentration range. In the system A/C an additional induced S_B phase occurs between 45 and 85 mole% C. In the system A/B an intermediate S_C phase (60—75 mole% B) is observed.

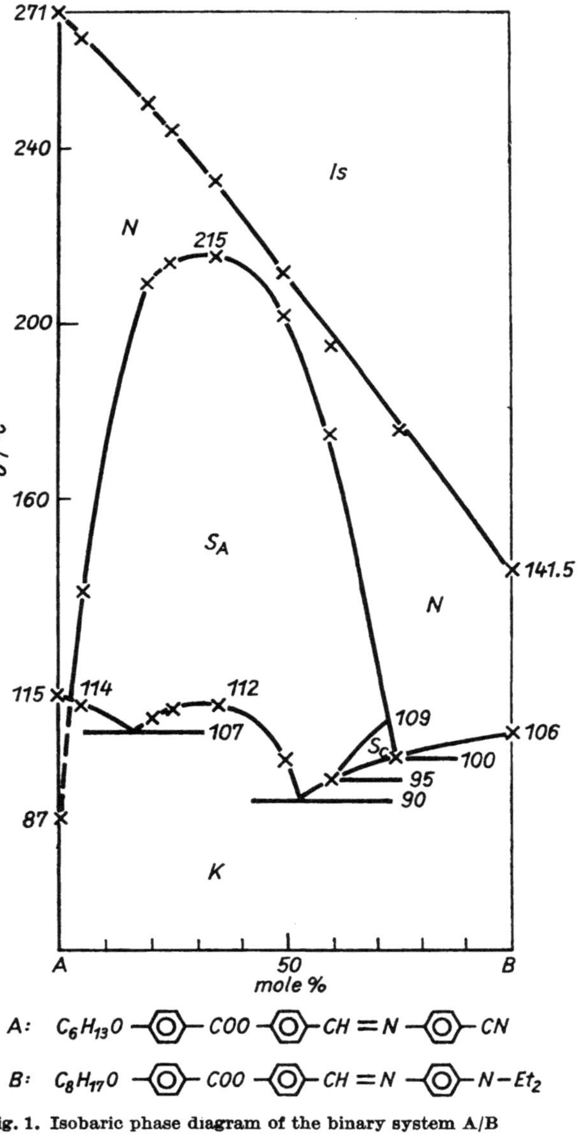

Fig. 1. Isobaric phase diagram of the binary system A/B

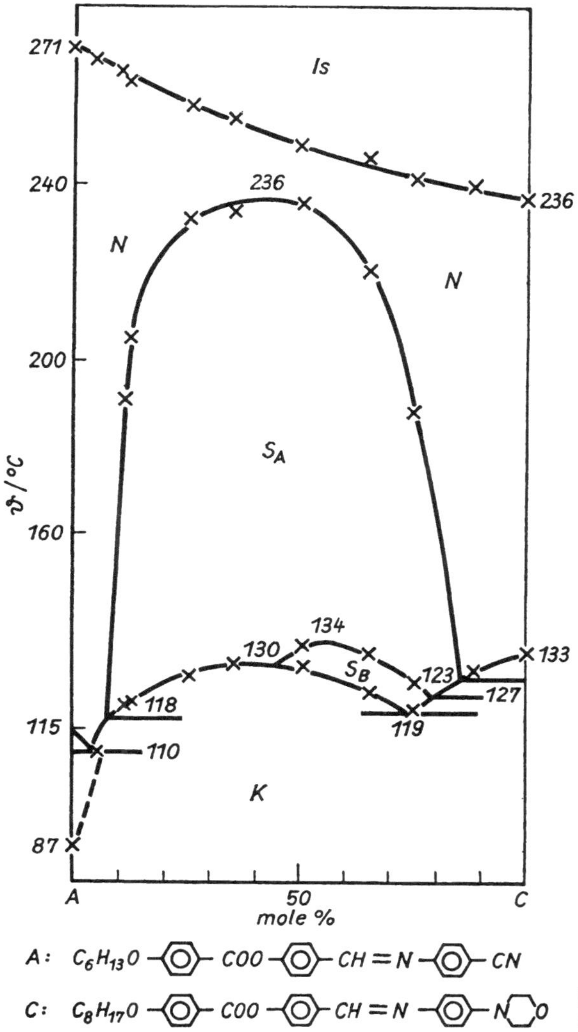

Fig. 2. Isobaric phase diagram of the binary system A/C

The complex formation in the system A/C is indicated by a strong yellow-brown colour of the mixed phases, whereas the single compounds A and C are colourless or weakly yellow, respectively. Although in the system A/B a direct evidence of complexing was not possible, EDA interaction is probable because the dialkylamino group of compound B generally is a stronger electron donator than the morpholino ring of compound C (MUSTROPH, BARTEL).

Possibly in the system A/B the charge transfer band is covered by the broad long wave band of compound B. On the other hand, the maximum of the melting curve is an important hint for complexing in both binary systems indicating complex formation in the solid state.

The reason for the extreme enhancement of the S_A phase region could be that the three ring compounds used in our systems exhibit a stronger EDA interaction than comparable two ring compounds.

References

ARAYA, K., MATSUNAGA, Y.: Mol. Cryst. Liq. Cryst. **67**, 153 (1981); Bull. Chem. Soc. Japan **53**, 3079 (1980); **54**, 2430 (1981)

DE JEU, W. H., LONGA, L., DEMUS, D.: J. Chem. Phys. **84**, 6410 (1986)

DEMUS, D., PELZL, G., SHARMA, N. K., WEISSFLOG, W.: Mol. Cryst. Liq. Cryst. **76**, 241 (1981)

KOFLER, L., KOFLER, A.: Thermo-Mikromethoden zur Kennzeichnung organischer Stoffe und Stoffgemische, Weinheim 1954

MUSTROPH, H., BARTEL, R.: J. prakt. Chem. **324**, 743 (1982)

SHARMA, N. K., PELZL, G., DEMUS, D., WEISSFLOG, W.: Z. phys. Chem. (Leipzig) **261**, 579 (1980)

WEISSFLOG, W., PELZL, G., DEMUS, D.: Cryst. Res. Technol. **21**, 117 (1986)

WEISSFLOG, W., PELZL, G., WIEGELEBEN, A., DEMUS, D.: Mol. Cryst. Liq. Cryst. Lett. **56**, 295 (1980)

(Received March 9, 1987)

Authors' addresses:

Dr. G. PELZL, Dipl.-Chem. M. NOVAK, Prof. Dr. habil. D. DEMUS
Forschungsabteilung der Sektion Chemie der
Martin-Luther-Universität Halle-Wittenberg
Mühlpforte 1
Halle
DDR-4020

Dr. W. WEISSFLOG
VEB Laborchemie Apolda, Betriebsteil Spezialchemie Leipzig
Abt. Forschung und Entwicklung
Elsteraue 9
Leipzig-Lütschena
DDR-7143

Cryst. Res. Technol. | 22 | 1987 | 8 | K129–K131

Ts. Z. Vitkina, E. M. Smolyarenko, V. M. Trukhan

Institute of Physics of Solids and Semiconductors of
the BSSR Academy of Sciences, Minsk, USSR

Magnetic Properties of Solid Solutions in the ZnP_2–CdP_2 Quasibinary System

The compounds of the $A^{II}B^{V}$ group have been studied extensively enough and their diagrams of melting have been plotted in the main. The phase diagrams of quasibinary sections formed by binary compounds are also available. In the $A_3^{II}B_2^{V}$–$A_3^{II}B_2^{V}$ systems continuous series of solid solutions are formed; in the $A^{II}B_2^{V}$–$A^{II}B_2^{V}$ systems the continuous series of solid solutions are formed by anion substitution, i.e. in ZnP_2–$ZnAs_2$ (Smolyarenko, Trukhan) and CdP_2–$CdAs_2$ systems (Marenkin et al.). The $ZnAs_2$–$CdAs_2$ system refers to the eutectic type (Marenkin et al.), whereas the ZnP_2–CdP_2 system — to the peritectic one (Smolyarenko, Trukhan). The phase diagram of the ZnP_2–CdP_2 system is given in Figure 1. The analysis of the phase diagram reveals that the boundary solid solution based on the tetragonal CdP_2 forms by means of the peritectic reaction $Lp + \alpha_B^{tB} \rightleftarrows \beta_a$, whereas the boundary solution α_b transforms into an intermediate γ-phase. In this case the nonvariant horizontal line b_1a_1 appears on the phase diagram in accordance with the phase rule. Thus three types of solid solutions are observed in the ZnP_2–CdP_2 system: boundary α-solid solution with a monoclinic structure, boundary β-solid solution with tetragonal structure and a large area of intermediate γ-solid solution.

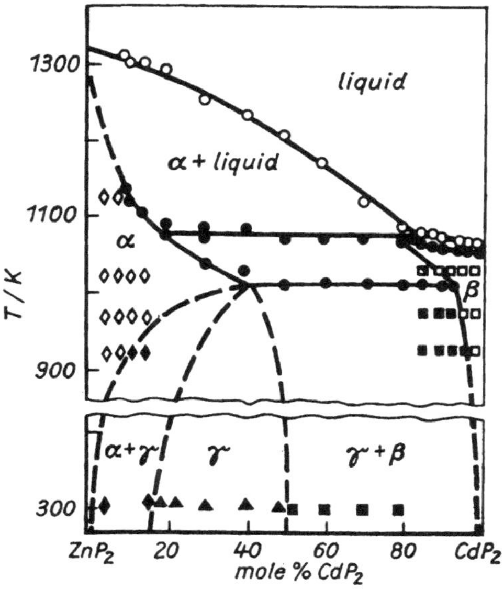

Fig. 1. Phase diagram for the $Z_xCd_{1-x}P_2$ system

It is established that the γ-phase has a monoclinic structure which is a derivative of the α-phase and forms as a result of ordering of zinc and cadmium atoms. In this case the doubling of the "c" parameter takes place. In comparison with the α-phase the X-ray patterns of the ordered solid solutions display a marked distribution of the intensities of the diffraction maxima and additional lines characteristic of the superstructure with doubled parameter "c" of the unit cell. At the point of the transition $\alpha_b - \gamma$ the composition obeys the simple stoichiometric ratio (3:2) at 1015 K.

As seen from the phase diagram the continuous series of solid solutions has not been formed despite the identity of the structure of the ZnP_2 and CdP_2 compounds. It is possible that the differences in the interatomic distances larger than those in the ZnP_2–$ZnAs_2$ system and wide differences in melting temperatures hinder the formation of solid solutions over the whole concentration range.

The magnetic susceptibility of the β-ZnP_2-based and tetragonal CdP_2-based solid solutions have been measured by us. Figure 2 illustrates the temperature trend of the magnetic susceptibility curve. The anomalies of the temperature trend of the magnetic susceptibility curve due to the existence of phase transitions (Fig. 2, curves 1, 6) (VITKINA, ORLIK et al.) are observed both in β–ZnP_2 and in CdP_2. The diagram of magnetic susceptibility as a function of temperature at $T = 370$ K for CdP_2-based alloy containing 0.07 mol. fract. of ZnP_2 displays a vivid jump of magnetic susceptibility. It seems that substitution of Cd by Zn in the solid solution causes a displacement in temperature of phase shift due to the step-wise ordering of the sublattice of metal atoms. As for the β–ZnP_2 -based solid solution the observed anomalies in the range 180—300 K can be due to the changes both in structure and in the contribution into the susceptibility of charge carriers and require further complex investigation.

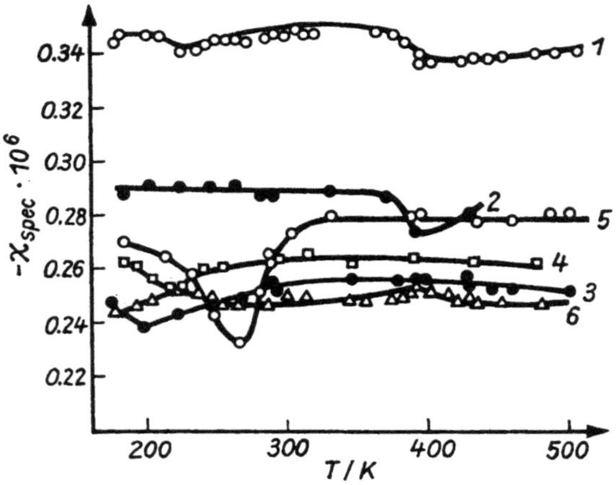

Fig. 2. Temperature trend of magnetic susceptibility for the $Z_xCd_{1-x}P_2$ solid solution system ($x = 0$ (1); 0.1 (2); 0.7 (3); 0.6 (4); 0.9 (5); 1 (6)).

Figure 3 gives the magnetic susceptibility as a function of concentration for the investigated alloys: the existing solid solutions based on zinc and cadmium diphosphides and those based on the compositions corresponding to the ordered γ-phase with a monoclinic structure. A clear break on the solubility curve of the $Z_xCd_{1-x}P_2$ system is apparent in the range $0.1 \leqq x \leqq 0.9$.

It should be noted that the substitution of cadmium by zinc in the range of the existence of solid solutions of tetrahedral modification decreases considerably the value of magnetic susceptibility which is indicative of the predominant change of the Langevin component of susceptibility. The fact that the molar magnetic susceptibility changes slightly testifies to the presence of the virtual deformation of electronic

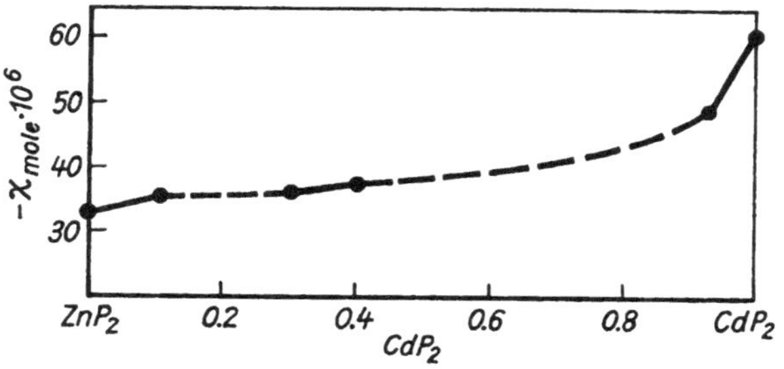

Fig. 3. Magnetic susceptibility as a function of concentration for the $Z_xCd_{1-x}P_2$ alloys

cloud and energy changes in the above mentioned alloys of monoclinic modification compensating the changes of the Langevin component of susceptibility connected with the changes of the mean radii of the atoms constituting the crystal.

References

VITKINA, Ts. Z., ORLIK, L. K., SMOLYARENKO, E. M., SHELEG, A. U., YAKIMOVICH, V. H.: Izv. AN BSSR, ser. fiz. N 5 (1985)

MARENKIN, S. F., KOVALEVA, I. S., SAIDULLAEVA, M., SANYGIN, V. P.: Izv. AN SSR. Neorg. mater. **19** (N 5) 837 (1983)

MARENKIN, S. F., MAKSIMOVA, S. I., PASHKOVA, O. N., SUKHORUKOVA, S. G., KHUSEINOV, B.: Zhurn. neorg. khimii **30**, N 11 (1985)

SMOLYARENKO, E. M., TRUKHAN, V. M.: Thermochim. Acta **95**, (1985)

(Received March 10, 1987)

Authors address:

c/o Ts. Z. VITKINA
Institute of Physics of Solids and Semiconductors
Academy of Sciences of the BSSR
ul. P. Brovki 17
220 726 Minsk, USSR

Cryst. Res. Technol.	22	1987	8	K132–K136

K. MELZER, Z. ŠIMŠA, M. ŁUKASIAK, J. SUWALSKI

Department of Physics, Karl-Marx-University, Leipzig, GDR, and
Institute of Physics, Praha, Czechosl. Acad. Sci., Czechoslovakia, and
Institute of Atomic Energy Swierk, Poland

Mössbauer Spectra and Electronic Properties of Titanomagnetite

Several Mössbauer studies of Ti-Fe spinels have been reported in the literature for $x \geqq 0.8$ (BANERJEE et al.; VANLEERBERGHE et al.) or for $x \leqq 0.05$ (UMEMURA et al.). BHADURI measured the Mössbauer spectra of the whole system ($0 \leqq x \leqq 1$) and suggested that $Ti_xFe_{3-x}O_4$ samples with $x < 0.5$ show the properties of delocalized electrons; whereas those with $x > 0.6$ may be regarded as n-type semiconductors with localized electrons.

We present measurements of electrical properties and Mössbauer spectra of synthetic $Ti_xFe_{3-x}O_4$ single crystals with $x = 0$; 0.24; 0.5; 0.72. It is seen that the electronic behaviour strongly depends on the content of titanium. At a composition of approximately $x = 0.65$ a transition from n- to p-type of electrical conductivity occurs. ^{57}Fe Mössbauer data of the samples for different temperatures obtained of the spectrum computer analysis using a least squares fit program are given. The Mössbauer spectra below and above the Néel temperature demonstrate the distinct decrease in mobility of electrons in dependence on the Ti content. It is evident from the Mössbauer spectrum of the sample $Ti_{0.72}Fe_{2.28}O_4$ recorded above the Néel temperature that Titanium is incorporated into the spinel structure of this system only in a four valent state.

Results and discussion

At $x = 0.24$ the room temperature Mössbauer spectrum does not deviate very much from that in Fe_3O_4. We can observe a tail on the inner side of each main peak. This deviation is stronger for $x = 0.5$ and is above all connected with the rise of the Fe^{2+} content at A sites. In the case of $x = 0.72$ the room temperature spectrum already shows a transition from the ferrimagnetic to the paramagnetic state. The results of the spectrum computer analysis are listed in Table 1.

The first subspectrum (and subspectrum 2 for $x = 0.5$) is related to Fe^{3+}(A). The second ($x = 0$; 0.24) or the third ($x = 0.5$) subspectrum is related to $Fe^{\approx 2.5+}$(B). This implies that at least in the case of $x \leqq 0.5$ rapid electron hopping takes place at B sites. The third ($x = 0.24$) or the fourth ($x = 0.5$) subspectrum is related to iron at A and B sites with a two-valent character.

Three models of the cation distribution of $Ti_xFe_{3-x}O_4$ were proposed (BANERJEE et al.). The subspectrum intensities agree well with the AKIMOTO model. But we must consider the strong sensitivity of the spectrum intensity to the fit parameters chosen. Some intermediary situation between AKIMOTO and O'REILLY-BANERJEE models seems to us the most probable (ŠIMŠA et al.).

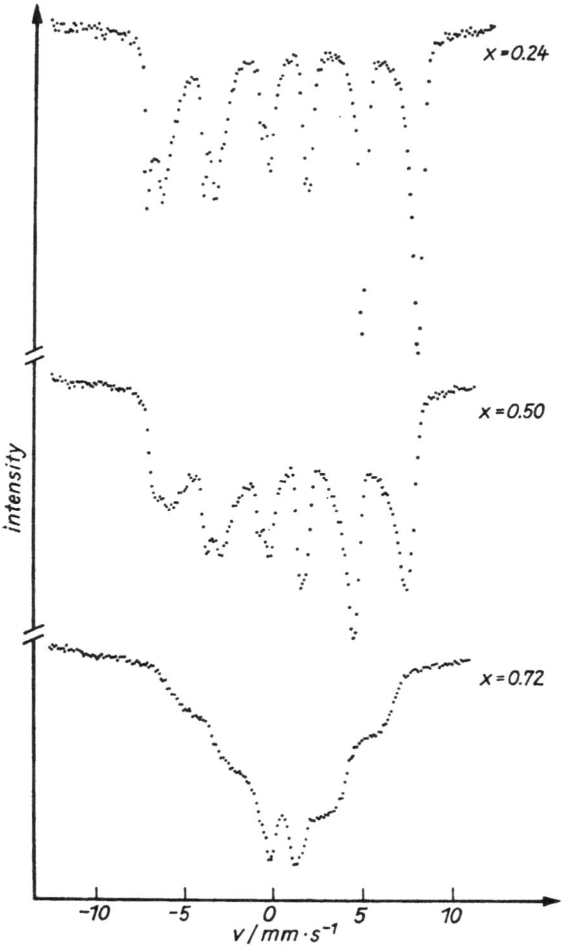

Fig. 1. Room temperature Mössbauer spectra of the samples investigated

Table 1

Calculated room temperature Mössbauer parameters of the $Ti_xFe_{3-x}O_4$ system

	$x = 0$		$x = 0.24$		
subspectrum	1	2	1	2	3
intensity (%)	33	67	23	36	41
isomer shift (mm/s)	0.29	0.69	0.29	0.62	0.73
H_{eff} (kOe)	489	460	480	447	398

	$x = 0.5$			
subspectrum	1	2	3	4
intensity (%)	13	3.5	34	49.5
isomer shift (mm/s)	0.31	0.15	0.67	0.79
H_{eff} (kOe)	447	441	398	278

(isomer shift relative to iron metal)

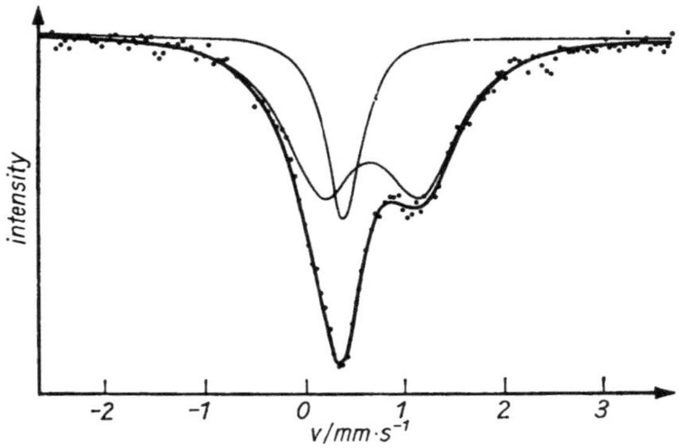

Fig. 2. Mössbauer spectrum for $x = 0.72$ recorded at $T = 423$ K

Table 2

Calculated Mössbauer parameters of the sample $Ti_{0.72}Fe_{2.28}O_4$ at 423 K

subspectrum	1	2
intensity (%)	24	76
isomer shift (mm/s)	0.37	0.65
quadrupole splitting (mm/s)	0.06	1.01
linewidth (mm/s)	0.43	0.91

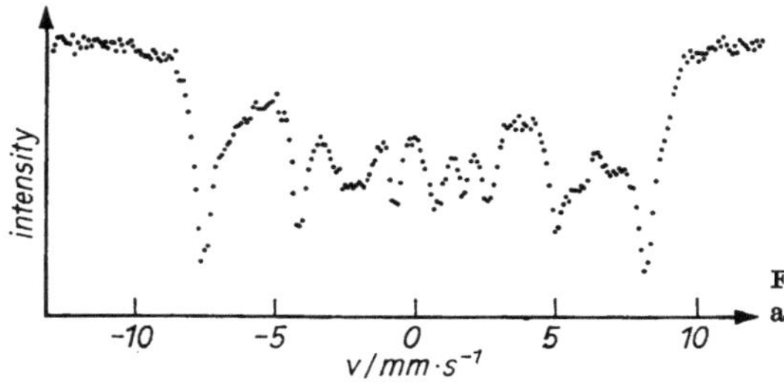

Fig. 3. Mössbauer spectrum for $x = 0.72$ recorded at $T = 4.2$ K

Table 3

Calculated Mössbauer parameters of the sample $Ti_{0.72}Fe_{2.28}O_4$ at 4.2 K

subspectrum	1	2	3	4 (doublet)
intensity (%)	39	20.5	34	6.5
isomer shift (mm/s)	0.42	1.2	1.26	1.65
quadrupole splitting (mm/s)	−0.08	−1.6	1.2	1.9
H_{eff} (kOe)	494	418	281	—

Above the Néel temperature the Mössbauer spectrum of Fe_3O_4 is very simple (LU-SAN PAN et al.). Only a broadened single line can be seen. The absence of any structure is remarkable. It is assumed that an electron exchange takes place between Fe^{2+} and Fe^{3+} in the octahedral sublattice, in the tetrahedral sublattice and in the two different sublattices (DIECKMANN et al.). As distinct from magnetite the Mössbauer spectrum of the sample with $x = 0.72$ recorded at 423 K (Néel temperature is 352 K) comprises two broad peaks of unequal intensities. (Fig. 2.) The computer fit results for two paramagnetic doublets are listed in Table 2. The calculated parameters cor-

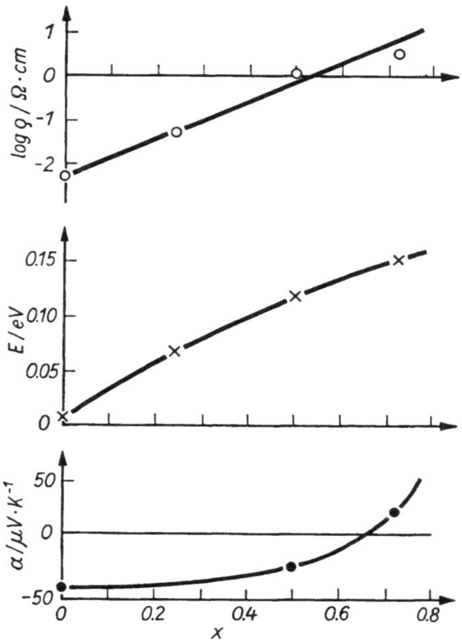

Fig. 4. Room temperature values of the electrical transport coefficients of $Ti_xFe_{3-x}O_4$

respond well to Fe^{3+} and to Fe^{2+}, respectively. The ratio of the intensities is in good agreement with the theoretical Fe^{3+}/Fe^{2+} ratio of the composition $Ti^{4+}_{0.72}Fe^{3+}_{0.56}Fe^{2+}_{1.72}O_4$. This result is additional evidence for the exclusive presence of Ti^{4+} in the samples. As can be seen, the lines are broadened. As VANLEERBERGHE discussed for samples with $x \geqq 0.9$ at least the Fe^{2+} doublet can be described as a superposition of four doublets for different A and B sites.

Mössbauer spectra of the sample with $x = 0.72$ were measured at 80 K and at 4.2 K. Both spectra are similar in shape but complex in structure compared with those at and above room temperature. Provisional computer fit results for the 4.2 K spectrum are summarized in Table 3. In the spectra there are three different sextets and one doublet. About 25% of the total area is related to Fe^{3+} contained in the first subspectrum. According to Table 3 four Fe^{2+} groups with different Ti^{4+} surrounding are observed. About one third of the subspectrum 1 area and the total area of the other subspectra are related to Fe^{2+}. The occurrence of a doublet caused by magnetically separated Fe^{2+} is remarkable.

The room temperature values of the electrical transport coefficients of the samples studied are shown in Figure 4. They include the logarithm of the electrical resistivity ($\log \varrho$), the activation energy E as determined from $\varrho = \varrho_0 \, e^{E/kT}$ and the Seebeck coefficient α (thermopower per 1 degree K). It is seen that at a composition of approximately $x = 0.65$ a transition from n- to p-type of electrical conductivity occurs which means that for the Ti-rich part of the system hole conduction prevails over electron transport. This result agrees well with the Mössbauer spectrum at 423 K of the sample with $x = 0.72$. In this case no electron exchange occurs. A simultaneous presence of both transport mechanisms hinders the exact evaluation of the parameters involved in ϱ_0. However, as the magnitude of ϱ_0 (which is proportional to τ, i.e., the time between two successive jumps or hops of electrons) does not change substantially ($\log \varrho_0 = -2.5$ for $x = 0$ and 0.24 and $\log \varrho_0 = -2$ for $x = 0.5$ and 0.72), we believe that rapid electron (or hole) hopping is the prevailing mechanism in the whole system studied. Moreover, due to the larger distances between A site cations, a contribution of tetrahedral hopping to the total conduction seems negligible to us except at temperatures above the Néel temperature.

The authors would like to thank Dr. V. A. M. Brabers, Eindhoven University of Technology, for the sample preparation.

References

Banerjee, S. K., O'Reilly, W., Gibb, T. C., Greenwood, N. N.: J. Phys. Chem. Solids **28**, 1323 (1967)
Bhaluri, M.: J. Chem. Phys. **77**, 1400 (1982)
Dieckmann, R., Witt, C. A., Mason, T. O.: Ber. Bunsenges. Phys. Chem. **87**, 495 (1983)
Lu-San Pan, Evans, B. J.: AIP CONF. Proc. **34**, 181 (1976)
Šimša, Z., Široky, P., Brabers, V. A. M.: Czech. J. Phys. **B30**, 957 (1980)
Umemura, S., Iida, S.: J. Phys. Soc. Japan **44**, 341 (1978)
Vanleerberghe, R., De Grave, E., Vanderberghe, R. E., Robbrecht, G. G.: J. Physique **41**, C1-179 (1980)
Vanleerberghe, R., Vandenberghe, R. E.: Hyp. Int. **23**, 75 (1985)

(Received March 14, 1987)

Authors' addresses:

Dr. K. Melzer
Sektion Physik der Karl-Marx-Universität Leipzig
Linnéstr. 5
Leipzig
DDR-7010

Dr. Z. Šimša
Fyzikàlni ústav, Československà Akademie VĚD
18040 Praha 8
Na Slovance 2

Doz. Dr. J. Suwalski, Dipl. Phys. M. Łukasiak
Instytut Energii Atomowej Swierk,
05-400 Swierk, Poland

Cryst. Res. Technol.	22	1987	8	K137–K140

E. M. Kostova, M. K. Kostov

High Institute for Civil Engineering and
Architecture, Department of Physics, Sofia, Bulgaria

Experimental Study of Optical Activity, Electrogyratory and Electrooptic Effects in $Bi_{12}GeO_{20}$

$Bi_{12}GeO_{20}$ and $Bi_{12}SiO_{20}$ are sellinite crystals of cubic symmetry — point group 23 (Abrahams et al.). These materials are attractive with their interesting physical properties: high photosensitivity (Lipson, Nisenson), natural optical activity (Lenzo et al.), electrooptic effect (Lenzo et al.) and electrogyratory effect (Lenzo et al.; Miller). The crystals are large and with excellent optical quality.

The natural optical activity, the electrooptic and the electrogyratory effects are reported by different authors using different specimens and different methods of investigations. The last two effects frequently have been taken as one (Lenzo et al.).

In this work we report on the experimental results for $Bi_{12}GeO_{20}$ about these three effects, using three different setups and three different methods. The experiments were carried out on the same specimen, possessing the three effects.

A (110) cut is suitable for measuring the electrooptic effect in a longitudinal configuration (the measurements will not be influenced by the electrogyratory effect). A (111) cut will give in a longitudinal configuration only the electrogyratory effect (Nye; Sirotin, Shaskolskaya). On the other hand the natural optical activity will be present in every cut.

In order to measure the three effects in one and the same specimen, a cut was chosen, possessing the three effects. The main purpose of this work was to separate the effects and estimate their magnitudes.

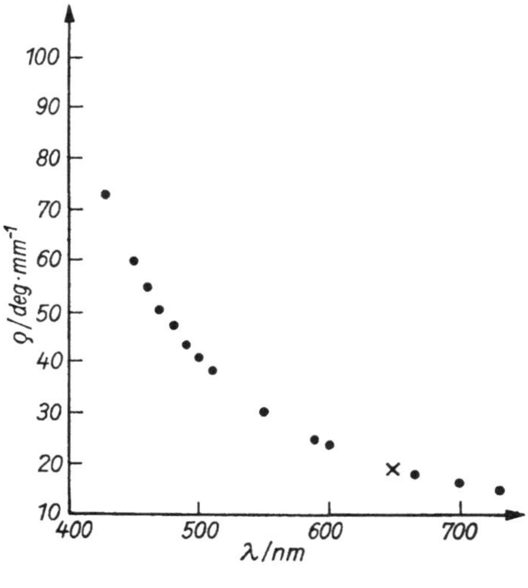

Fig. 1. Dispersion of the specific rotation of the plane of polarization for $Bi_{12}GeO_{20}$

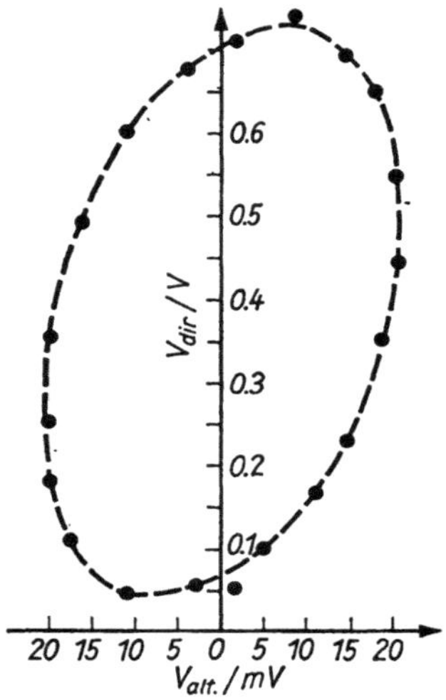

Fig. 2. Relationship between V_{dir} and V_{alt} for $Bi_{12}GeO_{20}$, $d = 0.985$ mm, $V_{rms} = 975$ V

The natural optical activity (the rotation of the plane of polarization of light, when the crystal is not subjected to external influences) was measured with a modulation polarimeter, working with an electrogyratory modulator (KOSTOVA, KOSTOV). In Figure 1 the dispersion of the specific rotation for $Bi_{12}GeO_{20}$ is given.

For measuring the electrooptic and electrogyratory effects a longitudinal configuration was used (the plane polarized light and the applied electric field were both along the big face of the crystal).

The electrogyratory effect was measured by a method (MILLER), which excludes the electrooptic effect. An alternating electric field at frequency of 565 Hz was applied. The experimental results are shown in Figure 2. From this figure for $Bi_{12}GeO_{20}$ of thickness $d = 0.985$ mm, applied electric field $V_{rms} = 975$ V, $\lambda = 650$ nm, $n = 2.52$ and using the formulae of MILLER for the electrogyratory coefficient γ we obtain

$$\gamma = 32.11 \cdot 10^{-13} \text{ m/V} .$$

The corresponding specific rotation ϱ_2 and angle of rotation φ_2 of the plane of polarization are

$$\varrho_2 = \frac{\pi}{\lambda n_0 d} \gamma \cdot V_{rms} = 8.5 \frac{\text{rad}}{m} \text{ (for the electrogyratory effect)}$$

$$\varphi_2 = \varrho_2 \quad d = 0.00837 \text{ rad} .$$

From Figure 1 for $\lambda = 650$ nm the specific rotation ϱ_1 due to the natural optical activity and the corresponding angle rotation φ_1 are

$$\varrho_1 = 19.45 \text{ deg/mm} , \qquad \varphi_1 = 19.45 \cdot 0.985 \text{ deg} = 0.33420 \text{ rad}$$

(for the natural optical activity) .

The total angle of rotation of the plane of polarization due to the natural optical activity and the electrogyratory effect is

$$\varphi_1 + \varphi_2 = 0.34257 \text{ rad}$$

and

$$[2(\varphi_1 + \varphi_2)]^2 = 0.46943 \text{ rad}^2 .$$

The electrooptic effect was studied by a recently developed method (HENRY et al.), which in the studied cut gives the combined effect: electrooptic + electrogyratory effects. Then using the magnitude of the electrogyratory effect, the electrooptic coefficient is calculated.

Two eliptically polarized waves propagate in an optically active and birefringent crystal (NYE). These two waves have the same ellipticity, opposite sign of rotation and perpendicular major axes, which coincide with the principal directions of vibration, if the crystal is not optically active. The phase difference between these two elliptically polarized components is (NYE)

$$\Delta^2 + \delta^2 + (2\varrho)^2 = \delta^2 + [2(\varrho_1 + \varrho_2)]^2 ,$$

where δ is the phase difference due to the electrically induced birefringence (the electrooptic effect), ϱ_1 is the phase difference due to the natural optical activity (the natural circular birefringence) and ϱ_2 — the phase difference due to the electrogyratory effect (the electrically induced circular birefringence).

For calculating δ the following experiment was caried out, Figure 3. For arbitrary and fixed position of the polarizer P, the analyzer A was rotated and the maximum and minimum intensities were measured and from these the function

$$V_\theta = \frac{I_{\max} - I_{\min}}{I_{\max} + I_{\min}} \quad \text{was calculated ,}$$

(see HENRY et al.). Then the polarizer was rotated through 45° and the measurements repeated for finding $V_{\theta+45°}$. In Table 1 the values of V_θ and $V_{\theta+45°}$ for different angles θ are given, differing by 10°. From the mean value of V_θ and $V_{\theta+45°}$ we found for $V_\theta^2 + V_{\theta+45°}^2 = 1.5218$. Using this value and the relation

$$\Delta^2 = \delta^2 + 0.469434$$

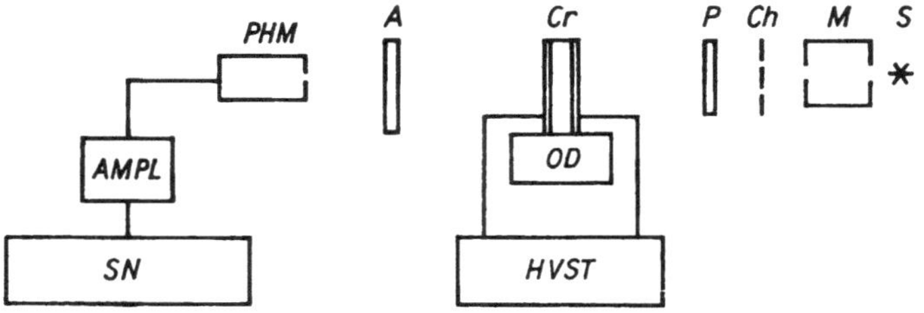

Fig. 3. Scheme of a setup for measuring the electrooptic effect: S source, M monochromator, Ch chopper, P polarizer, Cr crystal, OD orientation device, HVST high voltage stabilizer, PHM photomultiplier, AMPL amplifier, SN selective nanovoltmeter

Table 1

Angle θ/°	V_θ	$V_{\theta+45°}$
100	0.9746	0.7482
110	0.9419	0.8244
120	0.8574	0.8930
130	0.7825	0.9598
140	0.7279	0.9980
150	0.7774	0.9580
160	0.8274	0.9297
170	0.9103	0.8427
180	0.9624	0.7850

$\overline{V}_\theta = 0.8624$, $\overline{V}_{\theta+45°} = 0.8821$

from the equation

$$V_\theta^2 + V_{\theta+45^\circ}^2 = 1 + \left[1 - \frac{\delta^2}{2}\left(\frac{\sin\frac{\Delta}{2}}{\frac{\Delta}{2}}\right)^2\right]^2$$

we find

$$\sqrt{1 - \sqrt{1.5218 - 1}} = \frac{\delta}{\sqrt{2}}\left(\frac{\sin\frac{\sqrt{\delta^2 + 0.469434}}{2}}{\frac{\sqrt{\delta^2 + 0.469434}}{2}}\right).$$

From the last equation, using a suitable computer program δ was found. Then from

$$\delta = \frac{2\pi}{\lambda} n_0^3 r U$$

(where $U = 1360$ V is the applied voltage) the electrooptic coefficient r was found

$$r = (3.7 \pm 0.1) \cdot 10^{-12}\ \mathrm{m/V}\,.$$

Conclusion

Using three different setups and three different methods, the magnitudes of the three effects: natural optical activity, electrooptic effect, and electrogyratory effects are measured in one and the same specimen of $B_{12}GeO_{20}$ crystal.

References

ABRAHAMS, S. C., JAMICSON, P. B., BERUSTEIN, J. L.: J. Chem. Physics **47**, 4034 (1967)
HENRY, M., MALLIK, S., RONEDE, D.: J. Appl. Physics **59**, 2650 (1986)
KOSTOVA, E., KOSTOV, M.: Appl. Opt. **24**, 1726 (1985)
LENZO, P. V., SPENCER, E. G., BALLMANN, A. A.: Appl. Opt. **5**, 1688 (1966)
LIPSON, S. G., NISENSON, P.: Appl. Opt. **13**, 2052 (1974)
MILLER, A.: Phys. Rev. B. **8**, 5907 (1973)
NYE, J.: Physical properties of crystals, Moscow 1967 (transl.)
SIROTIN, J., SHASKOLSKAYA, M.: Principles of crystal physics, Moscow 1975, p. 646, Table D.14

(Received March 12, 1987)

Authors' address:

c/o M. K. KOSTOV
High Institute for Civil Engineering
and Architecture, Department of Physics
Boulvd Dragan Tsankov 2
1421 Sofia, Bulgaria

Continued from cover page 4

Short Notes

Crystal Research and Technology
Volume 22 **1987** **Number 8**

Contents

Original Papers

Continued on cover page 3

Crystal Research and Technology is indexed in Current Contents/PC & ES.

Bei Fragen zur Produktsicherheit wenden Sie sich bitte an:
If you have any questions regarding product safety,
please contact:

Walter de Gruyter GmbH
Genthiner Straße 13
10785 Berlin
productsafety@degruyterbrill.com